Engineering Analysis Using PAFEC Finite Element Software

Engineering Analysis Using PAFEC Finite Element Software

by

CHRIS H. WOODFORD
Lecturer in Computing

EVAN K.S. PASSARIS
Lecturer in Geotechnical Engineering

and

JOHN W. BULL
Lecturer in Civil Engineering
University of Newcastle upon Tyne

CRC Press
Taylor & Francis Group
Boca Raton London New York

CRC Press is an imprint of the
Taylor & Francis Group, an **informa** business

CRC Press
Taylor & Francis Group
6000 Broken Sound Parkway NW, Suite 300
Boca Raton, FL 33487-2742

First issued in paperback 2019

ISBN-13: 978-0-216-92901-2 (hbk)

Visit the Taylor & Francis Web site at
http://www.taylorandfrancis.com

and the CRC Press Web site at
http://www.crcpress.com

Preface

The aim of this book is to provide professional engineers and students of engineering with a sound working knowledge of the finite element method for engineering analysis and engineering design. For a number of years we have taught the finite element method with a view to its practical application. Our motivation in writing the present text was to produce a single readable text which would serve as a guide both to the method, and to its implementation in PAFEC (Program for Automatic Finite Element Calculations) software. At all times we were guided by the requirement to encourage good finite element practice and to develop the facility to handle genuine problems from the real world.

PAFEC software has gained world-wide acceptance and is used extensively by the UK academic community. Although this book concentrates on PAFEC software we do not regard this as a limitation: the software is generally available and even if other software is to be used, the book will still fulfil its major role in providing an introduction and guide to good finite element practice. The use of PAFEC-FE graphical facilities is explained and encouraged throughout the text. In the interests of clarity, drawings produced by this software are not reproduced in their entirety.

The book is divided into three sections. The introduction and chapter 1 present a predominantly non-mathematical guide to the history, development and underlying theory of the finite element method. The second section (comprising chapters 2 to 5) presents the detail of the implementation of the finite element method in PAFEC software: this second section will be of considerable help to the new user of PAFEC software. The final section (chapters 6 to 11) takes a selection of real problems and shows in step by step detail how they can be analysed. Real problems seldom have neat analytical solutions with which comparisons can be made, thus the engineering designer has to rely very much on acquired expertise when handling such problems and use the varying criteria available to decide if results produced by the computer are to be accepted or rejected. A major aim of this book is to enable the engineer to make those decisions and if necessary to modify the modelling process. Although problems are taken from a number of engineering disciplines, the presentations are deliberately general so that they will be of use to a wide spectrum of practitioners.

C.H.W.
E.K.S.P.
J.W.B.

Contents

Note: Introduction and chapters 1–5 by C.H. Woodford, chapters 6–8 by J.W. Bull and chapters 9–11 by E.K.S. Passaris.

Introduction

The finite element method

The finite element method as used in engineering analysis may be regarded as a modelling followed by an analysis. It is applied to structures which are sufficiently complex to deny immediate analysis. The geometry of the structure is modelled by a network or mesh of simple geometric shapes or elements, typically triangular or rectangular plates or solid blocks. The mathematical properties of the individual elements are designed to model accurately or as accurately as possible the physical characteristics they are required to represent. For example, a ship's hull might be modelled by a mesh of two-dimensional plate elements. The mathematics of each plate would be designed to simulate the behaviour of a steel plate, something which is perfectly feasible over a reasonably sized area. Having dissected the structure the analysis is performed on each individual element. The analysis uses the mathematics which are available for each element and takes account of the interaction between each element and its immediate neighbours, and of the boundary conditions on the structure. The results from the individual elements are combined to produce results which apply to the whole structure. To continue with the example of the ship's hull, the combination of results from the individual plates produces the deformations and stresses of the whole structure under prescribed loadings.

Historical background

There is nothing new about finite elements. The term itself was first coined in 1960 but the underlying ideas are much older. We shall see that the finite element method provides an essentially simple and unified approach to the whole area of engineering analysis.

Archimedes and the calculation of π

Archimedes, the Greek engineer and mathematician, used finite elements although he is rarely credited for having done so. Since the very early days it was realized by direct measure and by mathematics, that the circumference of a circle was directly proportional to the diameter. The constant of proportionality is now denoted by the Greek letter π (pi). For a long time pi was taken to be equal to 3 (I Kings 7:23) but it was largely because of Archimedes that a more accurate value became accepted.

Archimedes approximated the circumference of a circle by the edges of a polygon (Figure 1). In effect he replaced the unknown and unmeasurable, in this case the circumference, by objects he could measure exactly, namely straight lines. The straight lines were his finite elements. By taking polygons of relatively few sides Archimedes was able to settle on a value which on conversion to the decimal scale would be 3.14. To check the results thoroughly Archimedes experimented with polygons, both inside and outside the circle, with as many as 96 sides. By going to such extremes, he was in fact establishing good finite element practice. Within the physical limits imposed by his method, Archimedes was taking as many finite elements as were necessary to establish a pattern of results from which firm conclusions could be drawn. Furthermore and vitally important, these conclusions accorded with observation.

The method of approximating a circle by a polygon in fact can be traced back even further to the Greek sophist, Antiphon who lived 200 years before

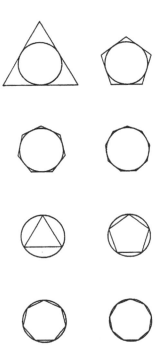

Figure 1 Approximation of the circumference of a circle by outscribed and inscribed polygons.

Archimedes. It may go back further still and demonstrates that in engineering, as in everything else, a good idea should be exploited to the full.

Newton and the integral calculus

Archimedes also considered areas as being made up of narrow rectangles. This idea was to be fundamental in the integral calculus formulated by Isaac Newton in the 17th century. Although, as illustrated in Figure 2, it may be intuitively obvious that any shape can be approximated by narrow rectangles, Newton gave the whole process a mathematical certainty. He showed that the area of the shape could be approximated by summing the areas of the narrow rectangles of equal width and that the degree of approximation increased as the number of rectangles increased. In theory total accuracy would be guaranteed with an infinite number of rectangles.

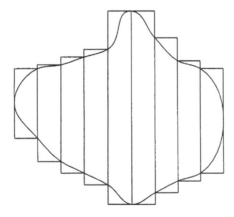

Figure 2 Approximation of a general shape by rectangles.

In practice there are limits to the number of rectangles which may be used to measure a given area. However, for any situation, it is found that to achieve a desired accuracy there is a limit beyond which nothing is gained by increasing the number of rectangles. This is analogous to the situation of Archimedes and his 96 sides, going further simply adds decimal places which are of no practical value. Once again the unknown has been measured in terms of less complicated items or what we refer to as finite elements. In the case just discussed the finite elements are the rectangles whose relevant properties are precisely known.

A number of formulae for calculating the areas under a curve may be known to readers. All are derived from the ideas above. The so-called 'trapezoidal rule' uses narrow trapeziums as the finite elements (Figure 3) whilst 'Simpson's Rule', carries the approximation a stage further by replac-

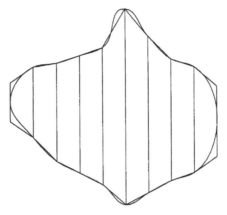

Figure 3   Approximation of a general shape by trapeziums.

ing the edges of the trapeziums by parabolic arcs. Simpson's elements are more complicated in their formulation but in general represent the original curve more accurately than trapeziums and so fewer of them are needed to achieve a prescribed accuracy. Whichever rule is used, it is applied repeatedly, each time increasing the density of the elements until the results stabilize to the required degree of accuracy.

Courant and the use of trial functions

In 1943 Courant introduced the idea of trial functions to simulate the behaviour of physical systems over small regions. Such physical systems might include the behaviour of a section of a shell or a beam or the pattern of heat flow through a medium. The word 'trial' is important in that it signifies a departure from the certainty of the modelling properties of the earlier finite elements we have mentioned. For example, in calculating the area under a curve using the trapezoidal rule there was a precision regarding the individual finite elements. The trapeziums modelled the small areas they covered exactly in the sense that the area could be formulated exactly (half the sum of the parallel sides times the distance between them). In moving to engineering applications we use functions, similar to Courant's trial functions, which can only approximate to the distortions and other phenomena which take place. It is in their very nature that mathematical functions model only an idealization of the real world. Nevertheless we are able to choose functions which give a fair measure of what is actually going on and are not too complicated to use.

Over the past few years many different types of finite elements have been devised for engineering applications. There are many because it is impossible to reflect every physical characteristic in just one or in even just a few elements. The mathematical complexities of providing a 'universal' element would be enormous. Different problems will generally require different elements. Throughout the book we will examine some of those that are currently available. For completeness it has to be said that a technique such as the finite element method is necessary for engineering analysis. In all but the very simple cases a structure would be too complex to analyse without some degree of approximation by composites of simpler structures.

Summary

Landmarks in the history of finite elements:

- 3rd century BC: Archimedes, measurement of the circle
- 17th century: Newton, the integral calculus
- 20th century: Courant, approximation by trial functions

Although it is beyond the scope of this book, it can be stated that there is a mathematical theory underlying the use of finite elements. The theory assures us that, as with Archimedes and his 96 sides the whole process of dissecting the model into smaller components, analysing those smaller components and combining all the results is sound. In general we can be confident that using more and more elements takes us closer and closer to the true solution of the problem as it has been presented for analysis.

However, it is important to realize that the results of the analysis may be useless and may indeed be positively dangerous if they are obtained from an idealization which does not realistically model the engineering structure under consideration.

Modern finite element practice

The ideas of Courant were first put into practice by aeronautical engineers in the 1940s. The development of jet aircraft produced a lot of structural problems for which reliable solutions were desperately required. From the 1940s the finite element method has been applied to practically every conceivable engineering problem. The applications described in later chapters are indicative of what is possible.

In the first stage of the analysis of an engineering problem the structure is represented in manageable form either on paper or on a computer. This modelling process is usually initiated by means of a rough drawing identifying the overall shape and dimensions of the structure together with an indication of boundary conditions which might be in the form of external loads, or applied pressures or prescribed temperatures. A note is made of the material properties of the components and some initial decisions are taken concerning the type of finite elements to be used and the manner in which they are to be assembled to model the structure. Clearly there are many ways in which the finite elements may be grouped or 'meshed'. Figure 4 illustrates how something as complex as an engine cylinder block might be modelled. Given the possible meshes and the different element types available, the scope for modelling an engineering structure is very wide indeed. Some models may yield meaningful results, others may not. An aim of this book is to guide the engineer towards making choices which are likely to produce reliable results.

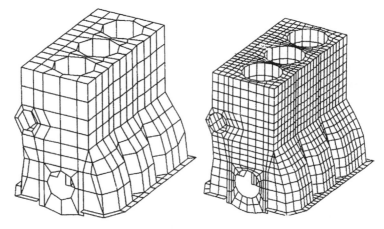

Figure 4 Engine cylinder block (reproduced by courtesy of NAFEMS, National Association for Finite Element Methods and Standards National Engineering Laboratory, East Kilbride, Scotland).

It must be emphasized that the modelling process is crucial. Even if apparently acceptable results are obtained it should always be the policy to examine the impact of variations in the model. If, as with Archimedes and his 96 sides the results stay very much the same despite using more and more elements it is reasonable to assume that what has been obtained is indeed representative of what is happening in the real world. The overriding consideration is whether or not the results accord with experience and the engineer's own intuitive feeling for what is right and what is wrong.

Summary

The steps in engineering analysis:

- Specification; what exactly is the problem
- Abstraction: formulation of a manageable model

- Analysis: mathematical analysis of the model
- Acceptance: acceptance or rejection of the results, with rejection implying a requirement for a more realistic abstraction

Inevitably the modelling process may involve considerable simplification particularly for larger structures. For example in modelling a ship or an oil rig we would consider the overall shape and not include every nut and bolt. In practice this is found to work well; the results obtained from finite element analysis conform to what actually happens. On the other hand the method is so powerful that it is possible to go down to the level of individual components so that nuts and bolts themselves may be analysed.

1 PAFEC-FE and the finite element method

1.1 The PAFEC-FE program

The aim of this chapter is to develop further an understanding of the finite element method. It will become apparent that the method is only viable if implemented in a computer program. It follows that a description of the method involves reference to its associated computer program. There are many commercially available programs for finite element analysis, but their diversity is such that it is necessary to concentrate on one particular product to develop a sound working knowledge. We have chosen PAFEC-FE, produced by PAFEC Ltd. and one of the market leaders.

1.1.1 Company background

PAFEC started in the mid-1960s. At that time the second generation of computers, the IBM 360s and the KDF9s, were becoming available to research workers. Given this quantum leap in computing power, attention was turned to methods of analysis which were previously well nigh impossible to implement for general use. In particular, attention was turned to the finite element method, which although conceptually quite simple, does require considerable computing resources. Interest in the finite element method has been further stimulated by the tumbling costs of computer hardware. In recent years the equivalent, or even better, of what once cost a million pounds may now be bought for a few hundred pounds. Not only have costs come tumbling down, but the computing power which once demanded a massive machine to be housed in a specially air-conditioned room and watched over by a team of operators and maintenance engineers can now be found sitting on a desk top.

Summary

Reasons for popularity of the finite element method:

- Can be used to solve a wide range of problems
- May be readily implemented on a computer
- Large number of complete computer systems (not only PAFEC-FE) now available

- Falling costs of computer hardware (though not computer software)

The advance of technology is ensuring that the power necessary for finite element analysis will be available to an ever increasing number of engineers. Among the early researchers into finite elements was a group in the Engineering Department of Nottingham University. For a while the practice was for individuals to write computer programs for individual use and although the results may have been communicated, the programs were not. This state of affairs led to a great deal of duplication of effort. A large part of any program for finite element analysis consists of sections for data input, mesh generation and routines for the solution of systems of linear equations. These sections are common to all finite element programs and once established and validated should be used by others. This was recognized by Richard Henshell who proposed that a common system be introduced at Nottingham. The system established was called PAFEC, Program for Automatic Finite Element Calculations. The system revolutionized finite element research. Researchers were freed from the task of writing computer programs to perform the mundane tasks of finite element analysis and were able to devote their energies to advancing the subject. In turn these advances were fed back into the PAFEC system for the benefit of the whole community.

Summary

Advantages of using a ready-made finite element computer system such as PAFEC-FE:

- The cumulation of many years of work is readily available
- Errors and inconsistencies will have been largely removed through user experience
- Such systems are written to be user friendly
- The user is free to devote full attention to engineering rather than the computer problems

Disadvantages:

- Results have to be taken on trust
- Some systems (though not PAFEC-FE) prevent the user from making changes to established procedures

By the mid-1970s the use of PAFEC had grown to such an extent that it was no longer possible to serve the ever increasing number of academic and industrial users from within the confines of a university department. In 1976 Richard Henshell and other members of the department left the University to form PAFEC Ltd. to market and develop the PAFEC program and associated services. Since then PAFEC Ltd. has grown to international status as a major supplier of engineering software. The range of products has increased but it is with the finite element software, the PAFEC-FE program as it is known, that we are primarily concerned.

1.2 Finite element analysis: a step by step guide

We are now in a position to identify the various stages in a finite element analysis of the model of an engineering structure. At each stage we will illustrate the generalities by means of a simple problem, namely the analysis of a cantilever truss. Whereas in practice the use of two- and three-dimensional elements predominates, our problem, which uses just one-dimensional elements, has the virtue of demonstrating the method without over complicating the mathematics.

1.2.1 Specification of the structure

The structure to be analysed is modelled by a discretization of elements having simple one-, two- or three-dimensional geometric shapes. To achieve this it is the practice to specify the overall model of the structure in terms of nodal points, or nodes which may be regarded as forming an outline. Each node is defined by its coordinates in some appropriate system and is uniquely identified by a number. Individual elements are then specified in terms of node numbers rather than coordinates. This makes the task much more manageable.

There will be occasions particularly if large, complex structures are involved, when many different discretizations or meshes will be possible. Chapter 4 gives some guidance on this but in the end it may be that the engineer has to be guided by experience. In any event experiments with different meshes should always be undertaken to try to establish a consistency of results.

Example. We consider a cantilever timber truss made up of five rods having equal cross-section and subject to an external load of 100 N (Figure 1.1). The rods will be assumed to be of uniform material of cross-section 0.25 m^2, having a Young's modulus of 9900 Pa, which corresponds to timber of strength class four. The choice of node numbers, element numbers and origin of axes is quite arbitrary and in no way affects the final outcome.

In this case, as in other simple examples, the formulation of the model follows quite naturally.

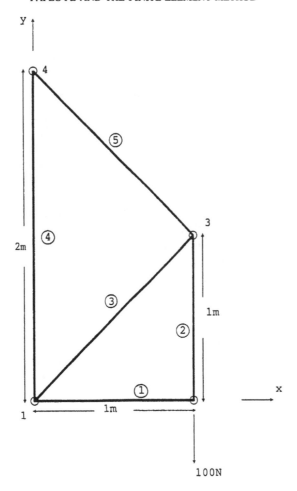

Figure 1.1 Cantilever truss idealization showing node numbers, element numbers (circled), axis set and the external load. Node 1 is only allowed to move in the y-direction. Node 4 is not allowed any movement.

1.2.2 Element formulation

In order to model a general configuration finite elements are generally simple one-, two- and three-dimensional shapes. In appearance they are straight lines, triangles or quadrilaterals with appropriate generalization to box and wedge shapes in three dimensions.

Whatever the shape we assume an equilibrium of forces and we are given a mathematical function which models physical properties.

The example (continued). In the case of the cantilever truss which we are going to analyse we will use a simple tension bar element (Figure 1.2). The

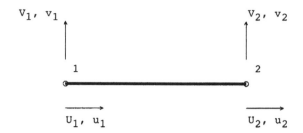

Figure 1.2   Tension bar U_1, V_1, U_2, V_2 are the forces acting in the directions shown at nodes
1 and 2, respectively. u_1, v_1, u_2, v_2 are the resulting displacements at similar positions.

element is simple in that uniaxial stress-strain conditions apply. We assume
that the bar element is in either tension or compression and so there is an
equilibrium of forces. This yields the first three equations of the set of linear
equations (1). The material property of the bar is reflected in the fourth
equation which describes the stress-strain relationship.

$$V_1 = 0, \quad V_2 = 0$$

$$U_1 = -U_2, \quad U_2 = \frac{-AE}{L}(u_1 - u_2) \tag{1}$$

where U_1, V_1, U_2, V_2 are the forces acting in the directions shown in Figure
1.2 at nodes (1) and (2), respectively. u_1, v_1, u_2, v_2 are the resulting displacements in the directions shown, A = area of cross-section of the
bar, L = length of the bar and E = Young's modulus.

These equations may be written in matrix form as

$$\frac{AE}{L}\begin{pmatrix} 1 & 0 & -1 & 0 \\ 0 & 0 & 0 & 0 \\ -1 & 0 & 1 & 0 \\ 0 & 0 & 0 & 0 \end{pmatrix}\begin{pmatrix} u_1 \\ v_1 \\ u_2 \\ v_2 \end{pmatrix} = \begin{pmatrix} U_1 \\ V_1 \\ U_2 \\ V_2 \end{pmatrix}$$

or $Kd = F$ where d is a column vector of displacements, F is a column
vector of forces and K is a matrix.

Traditionally K is termed the stiffness matrix, since the formula has all
the appearance of Hooke's law for springs. In apparently unrelated problems,
such as thermal problems where a similar form of equations arises, the term
stiffness matrix is still used.

In the case of the tension bar element, the linear form of the equations
arises quite naturally. In more complicated situations where physical properties might be represented by higher order equations including differential
equations, great ingenuity is used to linearize the equations. Inevitably this
approximation process involves a loss of accuracy but by working with small
elements it is hoped that errors are kept within reasonable bounds. This is

borne out in practice where it is found that by increasing the density of the mesh and therefore reducing the element size, the required accuracy is generally achieved.

The desire to obtain linear equations is prompted by the ease and certainty with which they may be solved by such guaranteed algorithms as Gaussian elimination. There is no correspondingly general theory for non-linear systems of equations.

1.2.3 Formulation of the element equations

Individual elements are described by linear equations of the type discussed in section 1.2.2. These equations will be relevant to some local axes system; for example in the case of the tension bar element the local Cartesian x-axis may be taken along the bar and the y-axis perpendicular. However, such an axes system may not take into account the position and orientation of the element within the overall structure. In order to have a single set of equations for the whole structure a global axis system is used. So where necessary, each set of local equations is transformed to equations in the global axis.

The example (continued). From Figure 1.1 it can be seen that element 1 lies along the global x-axis and so for this element at least the equations (1) derived above still stand. All the other elements lie along an axis which has been rotated relevant to the global x-axis and so for each of these elements the equations (1) have to be modified.

We consider a Cartesian coordinate system which has been formed by rotating an original Cartesian system through an angle (Figure 1.3). We know that a vector (x', y') in the new system is related to a vector (x, y) in the original system by the equations

$$x' = x \cos \theta + y \sin \theta$$
$$y' = -x \sin \theta + y \cos \theta$$

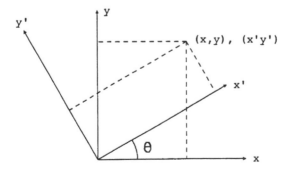

Figure 1.3 Rotation of axes.

To allow for a rotation factor the linear equations (1) are re-written in the dash notation and using the substitutions

$$U_1' = U_1 \cos \theta + V_1 \sin \theta$$
$$V_1' = -U_1 \sin \theta + V_1 \cos \theta$$

with similar expressions involving the pairs (U_2', V_2'), (u_1, v_1) and (u_2, v_2) yields

$$-U_1 \sin \theta + V_1 \cos \theta = 0$$
$$-U_2 \sin \theta + V_2 \cos \theta = 0$$
$$U_1 \cos \theta + V_1 \sin \theta = -U_2 \cos \theta - V_2 \sin \theta$$
$$U_2 \cos \theta + V_2 \sin \theta = \frac{-AE}{L} (u_1 \cos \theta + v_1 \sin \theta - u_2 \cos \theta - v_2 \sin \theta)$$

where all the forces and displacements are measured along axes in the global system and may vary from element to element. For example in the case of element 2, θ is 90°. In the case of element 1, θ is 0° and we revert, as expected, to the original form (1).

These equations may be written in matrix form as

$$\frac{AE}{L} \begin{pmatrix} 1 & 0 & -1 & 0 \\ 0 & 0 & 0 & 0 \\ -1 & 0 & 1 & 0 \\ 0 & 0 & 0 & 0 \end{pmatrix} \begin{pmatrix} \cos \theta & \sin \theta & 0 & 0 \\ -\sin \theta & \cos \theta & 0 & 0 \\ 0 & 0 & \cos \theta & \sin \theta \\ 0 & 0 & -\sin \theta & \cos \theta \end{pmatrix} \begin{pmatrix} u_1 \\ v_1 \\ u_2 \\ v_2 \end{pmatrix} = \begin{pmatrix} \cos \theta & \sin \theta & 0 & 0 \\ -\sin \theta & \cos \theta & 0 & 0 \\ 0 & 0 & \cos \theta & \sin \theta \\ 0 & 0 & -\sin \theta & \cos \theta \end{pmatrix} \begin{pmatrix} U_1 \\ V_1 \\ U_2 \\ V_2 \end{pmatrix}$$

Multiplying through by the inverse of

$$\begin{pmatrix} \cos \theta & \sin \theta & 0 & 0 \\ -\sin \theta & \cos \theta & 0 & 0 \\ 0 & 0 & \cos \theta & \sin \theta \\ 0 & 0 & -\sin \theta & \cos \theta \end{pmatrix}$$

namely

$$\begin{pmatrix} \cos \theta & -\sin \theta & 0 & 0 \\ \sin \theta & \cos \theta & 0 & 0 \\ 0 & 0 & \cos \theta & -\sin \theta \\ 0 & 0 & \sin \theta & \cos \theta \end{pmatrix}$$

yields the global stiffness equation

$$Ka = f$$

where a is a column vector of displacements measured in the global axis, f is a column vector of forces measured in the global axis and K is the matrix

$$\begin{pmatrix} \cos^2\theta & \cos\theta\sin\theta & -\cos^2\theta & -\cos\theta\sin\theta \\ \cos\theta\sin\theta & \sin^2\theta & -\cos\theta\sin\theta & -\sin^2\theta \\ -\cos^2\theta & -\cos\theta\sin\theta & \cos^2\theta & \cos\theta\sin\theta \\ -\cos\theta\sin\theta & -\sin^2\theta & \cos\theta\sin\theta & \sin^2\theta \end{pmatrix}$$

Using the notation u_i, v_i, U_i, V_i ($i = 1, \ldots, 5$) to denote displacements and forces at node number i measured in the global x and y directions, we can now formulate the equations in matrix form for each of the five elements. For the time being we will retain A and E in symbolic form and assume that we are working to SI units.

Element number 1: $\theta = 0$, $L = 1$ m

$$AE\begin{pmatrix} 1 & 0 & -1 & 0 \\ 0 & 0 & 0 & 0 \\ -1 & 0 & 1 & 0 \\ 0 & 0 & 0 & 0 \end{pmatrix}\begin{pmatrix} u_1 \\ v_1 \\ u_2 \\ v_2 \end{pmatrix} = \begin{pmatrix} U_1 \\ V_1 \\ U_2 \\ V_2 \end{pmatrix}$$

Element number 2: $\theta = 90°$, $L = 1$ m

$$AE\begin{pmatrix} 0 & 0 & 0 & 0 \\ 0 & 1 & 0 & -1 \\ 0 & 0 & 0 & 0 \\ 0 & -1 & 0 & 1 \end{pmatrix}\begin{pmatrix} u_2 \\ v_2 \\ u_3 \\ v_3 \end{pmatrix} = \begin{pmatrix} U_2 \\ V_2 \\ U_3 \\ V_3 \end{pmatrix}$$

Element number 3: $\theta = 45°$, $L = \sqrt{2}$ m

$$\frac{AE}{\sqrt{2}}\begin{pmatrix} \frac{1}{2} & \frac{1}{2} & -\frac{1}{2} & -\frac{1}{2} \\ \frac{1}{2} & \frac{1}{2} & -\frac{1}{2} & -\frac{1}{2} \\ -\frac{1}{2} & -\frac{1}{2} & \frac{1}{2} & \frac{1}{2} \\ -\frac{1}{2} & -\frac{1}{2} & \frac{1}{2} & \frac{1}{2} \end{pmatrix}\begin{pmatrix} u_1 \\ v_1 \\ u_3 \\ v_3 \end{pmatrix} = \begin{pmatrix} U_1 \\ V_1 \\ U_3 \\ V_3 \end{pmatrix}$$

Element number 4: $\theta = 90°$, $L = 2$ m

$$\frac{AE}{2}\begin{pmatrix} 0 & 0 & 0 & 0 \\ 0 & 1 & 0 & -1 \\ 0 & 0 & 0 & 0 \\ 0 & -1 & 0 & 1 \end{pmatrix}\begin{pmatrix} u_1 \\ v_1 \\ u_4 \\ v_4 \end{pmatrix} = \begin{pmatrix} U_1 \\ V_1 \\ U_4 \\ V_4 \end{pmatrix}$$

Element number 5: $\theta = 135°$, $L = \sqrt{2}$ m

$$\frac{AE}{\sqrt{2}}\begin{pmatrix} \frac{1}{2} & -\frac{1}{2} & -\frac{1}{2} & \frac{1}{2} \\ -\frac{1}{2} & \frac{1}{2} & \frac{1}{2} & -\frac{1}{2} \\ -\frac{1}{2} & \frac{1}{2} & \frac{1}{2} & -\frac{1}{2} \\ \frac{1}{2} & -\frac{1}{2} & -\frac{1}{2} & \frac{1}{2} \end{pmatrix}\begin{pmatrix} u_3 \\ v_3 \\ u_4 \\ v_4 \end{pmatrix} = \begin{pmatrix} U_3 \\ V_3 \\ U_4 \\ V_4 \end{pmatrix}$$

1.2.4 Assemblage of the equations

The individual equations formed in section 1.2.3 were derived by dissecting the structure into its component elements and considering each element in isolation. The assumption that there was an equilibrium of forces within each element is now extended to the whole structure. We also assume that the structure remains connected as originally specified. This enables us to combine the individual linear equations into a larger system of linear equations since the symbols for individual nodal forces and displacements retain their original meaning. At the same time the boundary conditions on the structure are included in the equations. These conditions may take the form of restraints on the structure, applied loads, prescribed displacements or in the case of thermal problems, prevailing temperatures.

The material properties of the structure have been included in the original equations and so we now have a full description of the model. In such cases the problem is said to be well defined and the corresponding set of linear equations has a unique solution.

If, however, the problem is not sufficiently well defined and as a result the structure is free to move, the system of equations will have an infinite number of solutions. In cases such as this PAFEC-FE software produces a warning message of which the following is an example:

WARNING ** FREEDOM   4 AT NODE   2 DIRECTION 2
LEADING DIAGONAL TERM   0.1818989E–11   CHANGED TO 1.0E20

For any analysis PAFEC-FE basically uses Gaussian elimination to solve the system of linear equations. Readers who are familiar with the method will know that in the case of systems having no unique solution the method breaks down. It becomes impossible to avoid a zero or near zero appearing on the diagonal of the system matrix as it is reduced to upper triangular form. In such cases the PAFEC-FE system replaces the zero or near zero (0.1818989E–11 in the example above) by an extremely large number (1.0E20) and so proceeds to a solution. In effect the corresponding unknown is constrained to take the value zero. If the unknown variable represents a displacement then the system is dictating that no movement is possible in a particular direction at that point. PAFEC-FE has added an extra constraint to the structure so that the analysis may proceed to a unique solution. In the example quoted above PAFEC-FE has added a constraint to node 2 in the 2 (or y-axis) direction which corresponds to degree of freedom number 4. It is for the user to decide if the program has made a physically meaningful decision. If not, then the original program data should be modified.

The example (continued). Continuing the analysis of the cantilever truss we note that the five sets of equations derived in section 1.2.3 cannot be solved separately since each set contains more unknowns than equations. However, by the principles of equilibrium and connectivity we are able to combine the equations into one large system of eight equations in eight unknowns, namely

$$AE \begin{bmatrix} 1+\frac{1}{2\sqrt{2}} & \frac{1}{2\sqrt{2}} & -1 & 0 & -\frac{1}{2\sqrt{2}} & -\frac{1}{2\sqrt{2}} & 0 & 0 \\ \frac{1}{2\sqrt{2}} & \frac{1}{2\sqrt{2}}+\frac{1}{2} & 0 & 0 & -\frac{1}{2\sqrt{2}} & -\frac{1}{2\sqrt{2}} & 0 & -\frac{1}{2} \\ -1 & 0 & 1 & 0 & 0 & 0 & 0 & 0 \\ 0 & 0 & 0 & 1 & 0 & -1 & 0 & 0 \\ -\frac{1}{2\sqrt{2}} & -\frac{1}{2\sqrt{2}} & 0 & 0 & \frac{1}{\sqrt{2}} & 0 & -\frac{1}{2\sqrt{2}} & \frac{1}{2\sqrt{2}} \\ -\frac{1}{2\sqrt{2}} & -\frac{1}{2\sqrt{2}} & 0 & -1 & 0 & 1+\frac{1}{\sqrt{2}} & \frac{1}{2\sqrt{2}} & -\frac{1}{2\sqrt{2}} \\ 0 & 0 & 0 & 0 & -\frac{1}{2\sqrt{2}} & \frac{1}{2\sqrt{2}} & \frac{1}{2\sqrt{2}} & -\frac{1}{2\sqrt{2}} \\ 0 & -\frac{1}{2} & 0 & 0 & \frac{1}{2\sqrt{2}} & -\frac{1}{2\sqrt{2}} & -\frac{1}{2\sqrt{2}} & 0 \end{bmatrix} \begin{bmatrix} u_1 \\ v_1 \\ u_2 \\ v_2 \\ u_3 \\ v_3 \\ u_4 \\ v_4 \end{bmatrix} = \begin{bmatrix} U_1 \\ V_1 \\ U_2 \\ V_2 \\ U_3 \\ V_3 \\ U_4 \\ V_4 \end{bmatrix}$$

We already know that u_1, u_4 and v_4 are zero so we can strike out the rows and columns corresponding to u_1, u_4 and v_4 from the system and so reduce the number of equations and unknowns from eight to five. In other words in forming the larger system we need only fill in rows and columns, other than those corresponding to u_1, u_4 and v_4. The equations $u_1 = 0$, $u_4 = 0$ and $v_4 = 0$ replace what was already there.

The system we obtain is now

$$AE \begin{bmatrix} \frac{1}{2\sqrt{2}}+\frac{1}{2} & 0 & 0 & -\frac{1}{2\sqrt{2}} & -\frac{1}{2\sqrt{2}} \\ 0 & 1 & 0 & 0 & 0 \\ 0 & 0 & 1 & 0 & -1 \\ -\frac{1}{2\sqrt{2}} & 0 & 0 & \frac{1}{\sqrt{2}} & 0 \\ -\frac{1}{2\sqrt{2}} & 0 & -1 & 0 & 1+\frac{1}{\sqrt{2}} \end{bmatrix} \begin{bmatrix} v_1 \\ u_2 \\ v_2 \\ u_3 \\ v_3 \end{bmatrix} = \begin{bmatrix} V_1 \\ U_2 \\ V_2 \\ U_3 \\ V_3 \end{bmatrix}$$

The quantities on the right-hand side are determined by the external loads on the structure. We have equilibrium and so $V_1 = 0$, $U_2 = 100$, $V_2 = 0$, $U_3 = 0$, $V_3 = 0$.

1.2.5 Solution of the equations

It is not difficult to realize that the number of equations to be solved can become very large. As more complex elements are used and their number increases it is not uncommon to be faced with the prospect of having to solve hundreds, thousands and indeed hundreds of thousands of equations.

We could, as we will in our example of the truss, simply solve the equations by standard Gaussian elimination storing the system matrix in the computer exactly as it appears on paper. However, this approach is unsuitable for anything but the simplest of problems since the storage demanded by the full system matrix would exceed the capacity of most modern computing systems.

A possible alternative is to re-order the equations so that the system matrix has a 'banded' form. In this case, only the non-zero part is stored in the computer. Judicious programming techniques which essentially factorize the matrix into the product of upper and lower triangular matrices may be used to produce a solution.

Considerable ingenuity has been employed to devise methods for solving the type of large systems of equations which occur in finite element problems. At all times the facilities and limitations of the available computer hardware have to be borne in mind. A currently popular technique is known as the Frontal Solution Method. This method is implemented in the PAFEC-FE system and consists of partially solving the system of equations as they are formed. The structure is analysed element by element and the unknowns corresponding to nodes which do not appear later in the process are replaced by expressions which involve other unknowns. In this way the size of the system which ultimately has to be solved is reduced. Once this system has been solved, steps can be re-traced to calculate the unknowns which were reduced out. For further discussion of this and other methods the reader is referred to the PAFEC-FE Theory Manual.

The example (continued). In our example we now give values to A and E namely 0.25 and 9900 and by conventional Gaussian elimination we obtain

$$v_1 = -0.04, \quad u_2 = 0.0, \quad v_2 = -0.12, \quad u_3 = -0.02, \quad v_3 = -0.08$$

We have assumed that we are working in SI units and so these calculated displacements are to be interpreted as fractions of a metre. At first glance we can say that the results appear to be reasonable.

The same problem may be solved using PAFEC-FE. Although at this stage the details will be unfamiliar, a suitable input to PAFEC-FE is listed in section 1.5. There will be relief all round that the results we have just obtained agree with those produced by PAFEC-FE.

1.2.6 Further calculations

Having found the displacements, it is then possible to calculate other quantities such as strains and stresses. Since these calculations are more or less routine and do not involve finite element principles we will not dwell on them here. Suffice it to say that in practice, whatever can be calculated will be calculated by the PAFEC-FE system with little or no prompting from the user.

1.3 Extension to plate and solid structures

The principles used in this chapter to analyse the construction of simple tension bar elements may be applied to more complicated structures. As we will see in subsequent chapters there are finite elements which model beam, plate, shell and solid structures. In all cases the set of linear equations defining the system is obtained in a comparable manner. It is assumed that there is an equilibrium of forces on the structure and that the structure remains connected. Material properties are reflected in the stress-strain relationships particular to the type of element being used. Specific examples will be quoted later. The inclusion of the boundary conditions completes the definition of the model. If the problem has been unambiguously defined we can expect the generated set of linear equations to have a unique solution corresponding to the solution of the problem.

1.4 Conclusion

Having identified the underlying principles and the major steps to be taken in a finite element analysis we now turn our attention in subsequent chapters to the implementation of those steps in a modern computer program, namely the PAFEC-FE system. We have already mentioned some features of the system but will now look at them in greater detail. We will also examine how to use the system in a manner to accord with good finite element practice.

1.5 PAFEC-FE data for the cantilever truss

```
TITLE Cantilever truss using the 34400 beam element.
CONTROL
C
C No special options required
C
CONTROL.END
C
C Nodal Coordinates – Origin at Node 1
C
NODES
   X      Y
0.00    0.00
1.00    0.00
1.00    1.00
0.00    2.00
C Beam positions
```

```
ELEMENTS
ELEMENT.TYPE=34400
PROPERTIES=11
TOPOLOGY
1  2
3  1
3  2
4  3
1  4
C
C Material properties
C
MATERIAL
MATERIAL.NUMBER  E
11                     9900
C
C Beam specifications
C
BEAMS
SECTION  MATERIAL  AREA
11        11        0.25
C
C Structural constraints
C
RESTRAINTS
NODE  PLANE  DIRECTION
  1     0       1
  4     0       12
  1     3       3
C
C External load
C
LOADS
NODE  DIRECTION  VALUE
  2      2        -100
END.OF.DATA
```

Reference

PAFEC Theory Manual, PAFEC Ltd., Strelley Hall, Strelley, Nottingham NG8 6PE.

2 PAFEC-FE—an overview

2.1 Modular nature of PAFEC-FE

The PAFEC-FE program has been of great assistance as a development tool in research, resulting in the advancement of the finite element method. This is largely due to the modular nature of the program. In effect the various components of the program are compartmentalized and only interact through well-defined procedures. Modularity permits new facilities to be added and existing facilities to be upgraded with relative ease.

The user of the PAFEC-FE program is in effect using a database with a database management system inherently provided. Within the database, information concerning a particular problem is stored, manipulated, updated and retrieved according to the rules of the mathematical subroutines performing the various types of engineering analysis required. At a more advanced level the modularity of the PAFEC-FE program permits the user to bend the existing rules and to introduce new ones.

A particular aspect of the modularity of the PAFEC-FE program will become apparent to the user through the way in which the various engineering characteristics of the problem to be solved may be specified in separate sections. The PAFEC-FE program accepts these sections, or data modules as they may be referred to, in any order and so the choice of ordering is left to the user.

Summary

Special features of PAFEC-FE:

- Modularity
- May be regarded as an engineering database

2.2 Specifying problem data

The advance of computer technology and in particular the development of peripheral devices has provided alternative methods for specifying the details of problems to be presented to the PAFEC-FE program for analysis. There is a trend towards interactive generation of the model of the engineering structure to be analysed. To this end, extensive use may be made of graphical display units, digitizing tablets, menus, mice and light pens to simplify and lighten the task of generating the model. PAFEC Ltd. have developed the PIGS (Pafec Interactive Graphics System) system for fully utilizing such equipment. Once the problem data has been produced it may then be submitted from PIGS to PAFEC-FE for analysis. Results may then be analysed using PIGS and if necessary modifications to the data may be made and the analysis repeated until an acceptable conclusion is reached.

The alternative more traditional approach consists of entering data into a computer disk file or data set. The prepared data set is then submitted to the PAFEC-FE program. Results can then be routed to a terminal, a printer or a graphics device as appropriate. Even though complicated structures may be generated more spectacularly perhaps using PIGS, the PAFEC-FE program using conventional disk files still has sufficient facilities to relieve the user of much tedious and repetitive effort. The use of disk files for the preparation of data is still the most widely used approach. For this reason it will be assumed unless otherwise stated that we are working solely with PAFEC-FE.

Summary

How data is supplied to PAFEC-FE:

- Directly from the keyboard, a mouse, light pen or digitizing tablet using PIGS
- Indirectly using a prepared data file

2.3 The ten phases of PAFEC-FE

The PAFEC-FE program is constructed as a number of phases to be executed in a manner progressing from pre-processor phases, through processor or solution phases and onto the final post-processor phases. These phases reflect the manner in which the PAFEC-FE program has implemented the various stages of finite element analysis as laid out in Chapter 1. Although not every phase will be required in every analysis the general progression through the system will be similar in most cases. A flow diagram is shown in Figure 2.1.

Phase 1: data input and verification. The pre-processor Phase 1, consists of the input and verification of data specifying the problem under consideration. The PAFEC-FE program attempts to identify any missing or conflicting data and in the event of detection of errors, issues mes-

Phase (1) Data input and verification
Phase (2) Automatic mesh generation
Phase (3) Data drawing
Phase (4) Further data generation
Phase (5) Further data drawing
Phase (6) Generation of the linear equations to be solved
Phase (7) Solution of the linear equations
Phase (8) Drawings of results
Phase (9) Stress calculations
Phase (10) Drawings of results

Figure 2.1 The ten phases of the PAFEC-FE program.

sages and prevents further execution so that the user may make appropriate corrections.

Phase 2: automatic mesh generation. For all but the very simple problems the generation of the finite element model is too tedious to be left to the user. The PAFEC-FE program has facilities for automating the procedure of constructing the computer model using finite elements as the building blocks. This process which is usually referred to as mesh generation takes place in Phase 2. During mesh generation the program will not allow any individual element to be distorted to an unacceptable extent. The levels of acceptability of element distortion have been determined by both mathematical analysis and user experience and are written into the PAFEC-FE program. In cases of severe distortion, error messages are issued and further execution prevented so that the mesh may be redesigned. In less extreme cases where distortion is contained within certain bounds, the PAFEC-FE program issues warning messages but permits execution to continue. These warning messages should not be disregarded. All subsequent results should be treated with extra caution. It has been found through bitter experience that the design of the mesh may critically affect the accuracy of the final results of the analysis. The whole meshing strategy is so crucial to successful finite element analysis that it is considered in more detail in a later chapter.

Phase 3: data drawing. The PAFEC-FE program provides facilities in this phase for viewing the generated finite element model from a number of aspects. Depending on what is available locally, drawings may be produced on either hard copy graph plotters or graphical display units. Figure 2.2 shows an example. In this case a plate is modelled by a four by four mesh of elements.

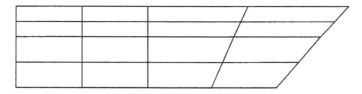

Figure 2.2   A PAFEC-FE data drawing.

Phase 4: further data generation. In this phase, the PAFEC-FE program sets up the boundary conditions of the problem as presented by the data. For example boundary conditions might arise from the specification of restraints on the structure to be analysed. Further, an edge or a surface might be constrained to be fixed rigid or permitted only to move in certain directions. As with mesh generation the PAFEC-FE program has facilities for simplifying the user's task of specifying restraints and other boundary conditions.

Phase 5: further data drawing. Phase 5 is similar to Phase 3, the only difference being that the boundary conditions evaluated in Phase 4 may now be incorporated and illustrated in the drawings. Phase 5 concludes the pre-processor stages. Successful completion of Phase 4 implies that the PAFEC-FE program is in a position to proceed to a possible solution of the problem with which it has been presented. Whether or not this problem is an accurate representation of the original problem, or what was thought to be the original problem, is quite another matter and one quite beyond the scope of any present day computer, and possibly any future computer, to decide. In addition there is no guarantee that the problem as represented is solvable. It may transpire that some hitherto undetected discrepancy in the data renders the problem either physically unrealistic or ambiguous. It may be that the problem as represented is so unstable as to defy the numerical techniques available, given the limited accuracy of modern computers.

Because of inherent computer 'round-off' error, solutions even to well defined problems may be so inaccurate as to be of no real value. These warnings apply not only to PAFEC-FE but to all computer programs for finite element or any other type of analysis.

Phase 6: generation of the linear equations to be solved. In this phase of the PAFEC-FE program the linear equations defining the individual properties of the individual elements in the finite element mesh and their relationship to their near neighbours are formulated.

Phase 7: solution of the linear equations. In Phase 7 the equations produced at Phase 6 are merged into a large system and are solved. If at this stage it became apparent that the number of unknowns is less than the number of equations a unique solution would not be possible. This would indicate that the original problem had not been completely specified. This implies there are 'loose ends' in the model of the structure. In other words sections of the structure are free to behave in an arbitrary manner without violating the original constraints. As explained in section 1.2.4, in order to be able to proceed to a solution the PAFEC-FE program imposes sufficient additional constraints to balance the number of unknowns and the number of equations. In a successful conclusion to this phase the results in the form of the values of the primary unknowns such as displacements and temperatures are tabulated.

Phase 8: drawings of results. A phase similar to Phases 3 and 5 which uses information from the solution phase, Phase 7. Figure 2.3 shows the plate of Phase 3 in both original and deformed shape.

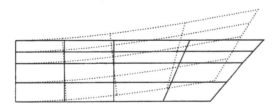

Figure 2.3   A PAFEC-FE drawing showing original and deformed shape.

Phase 9: stress calculations. Having obtained the primary unknowns in Phase 7, the PAFEC-FE program now proceeds to calculate the secondary unknowns, typically the stresses. The results are tabulated.

Phase 10: drawings of results. Another drawing phase for which the results from Phase 9 are used. Figure 2.4 shows the stress contours in the plate.

Figure 2.4 A PAFEC-FE drawing showing stress contours.

2.4 Types of analysis available

The PAFEC-FE program provides for various types of analysis. Often it will be the type of data modules present in the data set which will determine the type of analysis rather than any specific instruction in the control module.

The following is a list of the types of analysis available:

    (i) Static load analysis with displacement and stress calculations

   (ii) Thermal analysis with calculations of temperature distributions both transient and steady state and calculations of thermal stresses

  (iii) Dynamic analysis with calculations of natural frequencies and mode shapes; dynamic analyses of structure subject to prescribed forces including time dependent forces; seismic spectral response to earthquake ground motion

  (iv) Non-linear analysis; creep, plasticity and large displacement analyses

   (v) Solutions to acoustics, lubrication and piezoelectric problems.

2.5 The finite elements

It will be realized that to model effectively the shape of a general engineering structure finite elements of varying shapes and dimensions are required. The PAFEC-FE program provides a range of elements including one-dimensional beam elements, two-dimensional triangular and quadrilateral shapes and three-dimensional brick, wedge and tetrahedral shapes. There are even zero-dimensional elements in the form of mass and spring elements. There are shell elements which may be regarded as two-dimensional elements having a thickness rather than as three-dimensional elements.

For a given shape there is invariably a choice of elements dependent on the type of analysis required. The mathematical formulation of the element will have taken note of what is required. For example a two-dimensional element designed to have just two degrees of freedom in its own plane at each node could not be used where out of plane effects are expected. Again, an element for thermal modelling is likely to be very different in formulation from an element used in vibration analysis. In no sense is there a 'universal' element.

Given the number of possible shapes and the types of analysis available it is therefore not altogether surprising that the PAFEC-FE program provides more than a hundred different element types.

Summary

Finite element types provided by PAFEC-FE:

- Dimensions: zero, one, two and three
- Shapes: point, beam, triangular, quadrilateral, shell, brick and wedge
- Capabilities: stress, thermal and dynamic response applications

2.6 Running the PAFEC-FE program

Implementations of the PAFEC-FE program will inevitably vary from site to site, from computer to computer and from operating system to operating system. In general the user will initiate the PAFEC-FE program with a command such as PAFEC or PAFRUN followed by a name identifying the data set containing the problem data. A page is provided in Appendix A for the user to note local instructions. The user will require minimal expertise in the mechanics of the host operating system. The preparation of data and the listing of results will require some elementary file handling and editing skills.

2.7 Graphical output

The drawing facilities of the PAFEC-FE program are controlled through data presented by the user. Drawings of the structure both before and after analysis may be produced. A wide range of post-analysis drawings including drawings of displaced shapes and stress contours is available. Local conditions determine the hardware to be used for the drawings and this may include graphical display terminals and hard-copy plotters. Invariably there will be local facilities for producing plots which, including the PAFEC logo, can be of standard A4 size for direct inclusion in a report.

In the light of the calculations and the drawings produced by PAFEC-FE, modifications may be made to the original data and the analysis repeated. This may lead to yet further modifications with a consequent iterative cycle of analysis and modification until an acceptable solution is found; either that or the original problem has to be laid aside for fundamental remodelling.

2.8 PIGS

As has already been indicated the graphical capabilities of the PAFEC-FE program are greatly enhanced through use of the PIGS (Pafec Interactive Graphics System) system which is available as an optional extra. The user interacts directly with PIGS either through a mouse or through the keyboard at a graphical display terminal. The PIGS system can be used to generate a PAFEC-FE data set containing the finite element mesh with the applied loads and restraints. The user adds whatever other information is necessary to complete the model and to specify the type of analysis and results required. For example information relating to material properties might be added. The completed data set is then submitted to the PAFEC-FE program.

Drawings produced by the PAFEC-FE program may be relayed to the PIGS program and viewed in close-up and from any angle; a facility which is specially useful for viewing stress contours. The use of PIGS can simplify the generation of the original finite element mesh and permits extensive modification to regenerate existing models.

Whilst PIGS is an attractive addition, a knowledge of the PAFEC-FE system is still required. Whereas PIGS may simplify the mesh generation process and facilitate the interpretation of results it is still just a step in the direction of providing a full knowledge-based, artificial intelligent interface to PAFEC-FE.

Summary

Advantages of using PIGS with PAFEC-FE:

- Simplifies mesh generation
- Mistakes less likely
- Convenient for viewing drawings produced by PAFEC-FE

2.9 Will PAFEC-FE solve my problem?

Before even considering the use of the PAFEC-FE program or indeed any other program for analysis, the first step must be to define the problem clearly. From the outset the nature of the physical problem must be fully understood. Decisions have to be made as to how the problem is to be modelled and what is required from the analysis.

The next stage is to ensure that there are the PAFEC-FE data modules available to model every physical characteristic of the problem. The documentation of the types of analysis available through the PAFEC-FE program must be carefully studied to ensure that what is required is provided for and that all the relevant factors are taken into account. This is discussed further in chapter 5.

Finally, even if it appears that the PAFEC-FE program is an appropriate tool, the modelling process and the subsequent evaluation of the results must be undertaken with great care. The temptation to leave everything to the computer must be resisted. The PAFEC-FE program is a very powerful design tool and as such can be misused. The PAFEC-FE program has the capability to become an essential and effective design aid for the practising engineer and as such should be respected. Through practical examples from a wide range of applications, it is the purpose of this book to encourage and develop all the necessary familiarity and expertise.

Summary

Steps to be taken in using PAFEC-FE.

- Understand the problem; decide what is wanted
- Check that PAFEC-FE has the required capability
- Model and analyse using PAFEC-FE
- Check the results; re-model if not acceptable

3 The structure of the PAFEC-FE data set

3.1 Overview

The PAFEC-FE data set which appears at the end of chapter 1 is fairly representative and gives an idea of the format. In this chapter we examine the construction in detail.

In general a PAFEC-FE data set will contain a title, data modules and a control module and will be interspersed with user comments. A special end of data marker is used to terminate the data set.

The title is arbitrarily chosen by the user to classify the problem under analysis. The control module in addition to specifying options controlling the overall flow of the program may also specify the nature of the problem and the type of results required. The data modules specify the physical components of the problem including, for example, the finite element model of the structure, the materials used in the structure and the various loads and constraints upon the structure. With the exception of the end of data marker the various data sections may be arranged in any order. Figure 3.1 shows the structure of a typical PAFEC-FE data set.

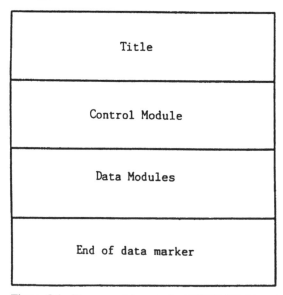

Figure 3.1 The general layout of a PAFEC-FE data set.

Summary

In a PAFEC-FE data set:

- The title record is optional, though recommended
- Data modules are essential in the sense that they are required to define the problem
- A control module is required
- Comment records are optional, though recommended
- The special end of data marker is obligatory

Difference in concept between a data module and a control module:

- In general, data modules define the problem and control modules control PAFEC-FE

3.2 Format conventions

Although the structure of data modules and control modules will be described in detail, the following points need to be borne in mind when constructing the records of a PAFEC-FE data set:

(i) All alphabetic characters of names special to PAFEC-FE are entered in upper case, i.e. as capital letters. Where PAFEC-FE allows names to be abbreviated to the first four characters, it follows that only the abbreviation need be upper case.

(ii) Data records must not exceed 80 characters, including blanks, in length. Longer data items may be continued onto following records. An asterisk (*) as the first non-blank character of a record indicates that the rest of the record is to be regarded as a continuation of the previous record.

At the other extreme two or more records may be compressed into a single record by means of the double slash sign (//). // indicates to the PAFEC-FE program that what follows is to be regarded as a new record. This facility may be useful in compressing data for listing and publication purposes.

(iii) The program skips over leading blanks at the beginning of a record.

(iv) Real numbers may be entered in either fixed point or exponential form or if appropriate as integers. 100.0, 1.0E+2 and 100 represent the same number to the PAFEC-FE program.

Data is entered as dimensionless quantities. PAFEC-FE is programmed to produce results using SI units and so users are recommended to enter data representing m, kg, N, s, J, W, etc. This consistency will ensure that results may be interpreted in the SI units implicit in the data. Other systems of units are possible but in such cases the user would be required to specify problem parameters such as material properties.

(v) Words and numbers are separated by either a comma (,) and/or by one or more blanks. The number zero, if it is the last complete number of a record may be omitted.

Summary

PAFEC-FE format conventions:

- Upper case for reserved words
- No more than 80 characters per data record
- A leading * is the continuation character
- Numbers follow usual computer language conventions
- Blanks and/or a comma act as separators

3.3 The title

The word TITLE at the beginning of a record indicates to the PAFEC-FE program that what follows is to be taken as a title for the problem defined by the ensuing data. The title will appear throughout the results produced by PAFEC-FE and so helps to identify the problem to the user and to others.

Two typical examples specifying titles to the PAFEC-FE program are shown here.

TITLE   Investigation of a Thin Shell.   J. Smith.

TITLE   Stress Calculation for Offshore Steel Structure
*   Marine Research Group - - - - - - - - - - - - - MRG/1/5/34

3.4 The data modules

The bulk of the data for a PAFEC-FE program is supplied in modular form by means of so-called data modules. Each data module describes an aspect of the model of the problem to be analysed. In a typical data set one module might define the nodal points of the structure, another might describe how the finite elements are meshed, another might describe the loads on the structure, another the restraints, and so on.

In some cases a module may refer to other modules. For example, the module describing the nodes has a facility to refer to a module defining an axes set. The axes module provides for definition of a coordinate system other than the default options of Cartesians. For some problems, for example applications involving cylindrical shells, it may be more convenient to define nodes in a polar coordinate system.

It is for the user to ensure that through the choice of appropriate modules the model is faithfully represented to the PAFEC-FE program. Although the specification of individual modules must conform to specified rules, the order in which they are arranged in the data set is optional.

The usual form of a PAFEC-FE data module is shown in Figure 3.2

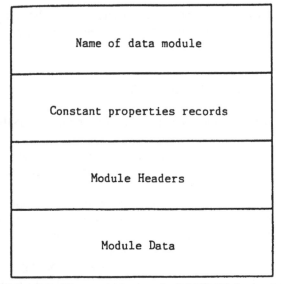

Figure 3.2 The general layout of a PAFEC-FE data module.

3.4.1 Module name

The first record of a data module always contains the name of the module. For example the module describing the nodes has the name NODES and similarly the module describing restraints on the structure has the name RESTRAINTS.

In all there are over a hundred modules available to users of the PAFEC-FE program, although any particular user within a given problem area will find that around ten are sufficient for most purposes. A classification is given in chapter 5.

In the PAFEC-FE program just the first four characters of module names are significant and so, for example, the abbreviations NODE and REST for NODES and RESTRAINTS, respectively, are acceptable.

3.4.2 Module headers

The various components of the data within a data module are grouped under specific names, or headers as they are referred to. The names of the available headers are specified in the definition of the module. For example the headers available in the NODES module include NODE.NUMBER, AXIS.NUMBER, X, Y and Z.

The data associated with NODE.NUMBER refer to the number assigned to a particular node. The corresponding information under AXIS.NUMBER, X, Y and Z defines the coordinate system and the position within that coordinate system.

The names of whichever headers are required may be supplied on the record immediately following the module name record. For example, the records

```
NODES
NODE.NUMBER AXIS.NUMBER  X  Y  Z
```

show the NODES module and the headers NODE.NUMBER, AXIS NUMBER, X, Y and Z.

There will be cases when not all the available headers are required. For example in a strictly two-dimensional problem there would be no need for the Z header in the NODES module.

The user may also wish to take advantage of the default options particular to the given header. If in the NODES module an AXIS.NUMBER is not specified then Cartesian axes are assumed. So specifying nodal positions in two-dimensional Cartesian coordinates would require no more than the NODE.NUMBER and X and Y headers. Similarly if NODE.NUMBERS is not specified then PAFEC-FE will designate the nodes to be in ascending order beginning with node number 1.

Finally there is a default option of specifying no headers at all. In this case the PAFEC-FE program assumes headers in the order in which they are specified in the definition of the data module.

Once again, as with most names in the PAFEC-FE program the user may abbreviate to the first four significant characters. So in the example previously quoted NODES, NODE.NUMBER and AXIS.NUMBER could be abbreviated to NODE, NODE and AXIS, respectively.

```
NODE
NODE  AXIS  X  Y  Z
```

3.4.3 Module data

Having either specified headers explicitly or having opted for the default headers, the actual module data is supplied on following records. Each module data record supplies items of information to the PAFEC-FE program. The order of the data within a module data record must correspond to the order laid down by the headers record.

As an example consider the specification of nodes at positions (1.0,1.5), (1.0,2.0), (1.0,2.5) in Cartesian coordinates and suppose that these nodes are to be given the node numbers 1, 2 and 3, respectively. Using the NODES module this could be specified to the PAFEC-FE program by the following:

```
NODES
NODE.NUMBER   X    Y
     1       1.0  1.5
     2       1.0  2.0
     3       1.0  2.5
```

3.4.4 Constant properties

In order to reduce the amount of data preparation so-called constant property records may be inserted immediately following the module name record and before the headers record (if any).

This facility is illustrated by considering the example above in which it will be noted that X takes the value 1.0 throughout the data module. The alternative form of the NODES module data shown below achieves the same result

```
NODES
X = 1.0
NODE.NUMBER   Y
     1       1.5
     2       2.0
     3       2.5
```

3.4.5 Repetition records

Yet further simplification of the preparation of data is available through use of the repetition facility. Again taking the previous example the generation of the three points having Cartesian coordinates (1.0, 1.5), (1.0, 2.0), (1.0,2.5) and nodal numbers 1, 2 and 3 could have been achieved using the following form of the NODES module:

```
NODES
NODE.NUMBER   X    Y
     1       1.0  1.5
   R2  1     0.0  0.5
```

The character string R2 indicates to the PAFEC-FE program that the previous record is to be repeated twice with the increments 1, 0.0 and 0.5 in the respective positions.

Although the examples given in this section have been quite trivial in the sense that no great saving of effort has been achieved it will be appreciated that for larger problems the facilities described may considerably lighten the task of data preparation.

Summary

Points to note in constructing data modules

- Specify the name of the module
- Identify constant properties
- Identify the headers required
- Use default headers and the implied default data where possible
- Identify possible repetition factors in the data

Finally, no special action is required to signify the end of a data module. The appearance of another data module, a control module, a title record or the special end of data record signifying the end of the complete data set is sufficient.

3.5 The control module

The control module provides controlling information to the PAFEC-FE program. This information may include details of how the flow of the PAFEC-FE program is to be controlled and details of the type of analysis to be performed. Other controlling information might determine such program parameters as the arithmetic accuracy to be used in the calculations and the disk and working store capacities to be made available.

Information to be supplied to the PAFEC-FE program through the control module is supplied in the form of keywords with sometimes an associated constant. Figure 3.3 shows the layout of a PAFEC-FE control module.

Control information is supplied on separate records between the opening and closing record. For example

```
CONTROL
PLASTIC
DOUBLE
CONTROL.END
```

would indicate that a plasticity analysis using double precision arithmetic was required.

Every PAFEC-FE data set must contain a CONTROL module. Even if the user wishes to take advantage of all the default options a null CONTROL module is supplied; that is a CONTROL module of the form

```
CONTROL
CONTROL.END
```

There are many control options in PAFEC.FE, a list is shown in Figure 3.4 together with an indication of their application. The next few sections illustrate the use of some of these options.

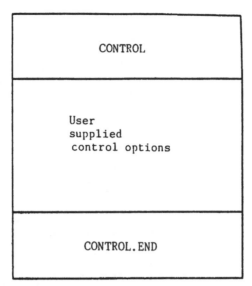

Figure 3.3 The general layout of a PAFEC-FE control module.

3.5.1 Phase control

Generally speaking the PAFEC-FE program will proceed through the various phases in a manner determined by the data. This default option may be overridden using FULL.CONTROL and PHASE records within a control module. In this case only those phases which are present on the PHASE record are executed.

The example below would cause only Phases 1, 2 and 3 to be run, and might be used if it were only required to generate and view the finite element mesh.

```
CONTROL
FULL.CONTROL
PHASE = 1, 2, 3
CONTROL.END
```

Control information relating to individual phases may be supplied on records following the appropriate phase record within a control section. For example, the phase control shown above could equally be achieved using the following control module:

```
CONTROL
PHASE = 3
STOP
CONTROL.END
```

In this case the process of automatically running through all the phases would be prematurely terminated at the end of Phase 3.

```
Execution control

FULL.CONTROL              Only phases named in the control option
                          PHASE= are executed.
STOP                      Stop execution after the current phase.
USE.                      Incorporate user FORTRAN into PAFEC-FE.
SKIP.COLLAPSE             Nodes within the tolerance are not
                          collapsed into a single node.
REDUCED.OUTPUT            The output for the current phase is
                          reduced to the essentials.
FULL.OUTPUT               Text output from the drawing phases is
                          not deleted.
CONCATENATE.OUTPUT        Output from all phases is written to a
                          single file.
SAVE   MORE.DATA          Range of options for using the save and
APPEND  DELETE            restart facility. Existing data may
REPLACE.MODULE            be modified at the restart.
STORE.INCREMENT=
SAVE.INCREMENT=
BASE=                     Size of BASE array for working space.
TOLERANCE=                Tolerance used in identifying coincident
                          nodes.
NON.LINEAR.TOL=           Stress/strain tolerance in plastic
                          analysis.
C.SLOPE=                  Plastic stress/strain constant.
GAP.ITERATION=            Maximum number of gap iterations.
```

```
Analysis control

AXISYMMETRIC              Structure is axisymmetric about the
                          global x-axis.
PLANE.STRAIN              Structure lies entirely in one plane.
STRESS                    In the absence of a STRESS.ELEMENT
                          module all elements are stressed.
BUCKLING                  Linear buckling analysis.
CREEP                     Creep analysis.
LARGE.DISPLACEMENTS       Non-linear large displacement analysis.
PLASTIC                   Plastic analysis.
SNAKES                    Non-linear static analysis.
LUBRICATION               Lubrication analysis.
CALC.STEADY.TEMPS         Range of options for thermal calculations
CALC.TRANS.TEMPS          including saving and restoring
SAVE.TEMPS.TO             temperatures for thermal stress analysis.
READ.TEMPS.FROM
```

Figure 3.4 Classification of the major PAFEC-FE control options.

3.5.2 Program estimates

The PAFEC-FE program is written in the FORTRAN language. A consequence is that the program is unable to allocate working storage areas
dynamically. This means that the size of the FORTRAN array it uses for
working store has to be known to the program at the outset of a particular
phase. There is a default value, currently set at 33 000 but for large problems
this will have to be overwritten by the user.

The PAFEC-FE program uses the terms 'BASE' and 'BASE SIZE' to
identify the name of the working array and its size. For example to allocate

a working array of size 50 000 for Phase 7 the following control section could be used:

```
CONTROL
PHASE = 7
BASE = 50000
CONTROL.END
```

The specification of the base size inevitably begs the question of how can an appropriate value be allocated before the program is run. Admittedly it is a difficult situation in that until the program has completed its execution it cannot be known for sure if a specified base size is adequate. Notice may be taken of the estimates produced by the PAFEC-FE program at the end of each phase. By the end of Phase 4 estimates for the rest of the phases are produced. Other than that either 'dummy' runs or user experience are useful guides.

At some sites other estimates may be required. These may include estimates of total computing time and estimates of the total program output requirement in terms perhaps of number of printed pages.

3.5.3 Save and re-start facilities

In order to allow a 'breathing space' between phases, to give time for further consideration of the problem in the light of intermediate results the PAFEC-FE program incorporates a save and re-start facility. This facility may also be used in cases where to run a complete program in one uninterrupted execution might be prohibitive in terms of local computing arrangements.

As an example, the following control options would cause the program to terminate execution after completion of Phase 7 and to store the results ready for a re-start at a later session.

```
CONTROL
PHASE = 7
STOP
SAVE
CONTROL.END
```

The program might then be re-started at Phase 8 using the following control section:

```
CONTROL
FULL.CONTROL
PHASE = 8, 9, 10
CONTROL.END
```

There would be no need to supply any more data, other than an END.OF.DATA record since all the relevant information would have been stored by the earlier run and the local system would allocate the appropriate files.

There are further facilities for modifying existing data using the MORE.DATA control option. For example, having generated and saved a mesh it is possible to supply new loading conditions without having to return to the very first phase.

In addition to the save and re-start facilities described here there are also facilities for saving and re-starting within Phase 7. Phase 7 is specially provided for since it is the solution phase and likely to be the most demanding in terms of computer processing time.

3.5.4   Modifying the PAFEC-FE program

Unlike many of its competitors PAFEC Ltd. does not deny users access to the FORTRAN source of PAFEC-FE. The company is to be congratulated on this enlightened approach. It does appear that they have not suffered commercially, the product being too large and too interwoven to be susceptible to piecemeal copying. In addition the product has benefited from the number of improvements that users having examined parts of the source have been able to suggest.

The control option USE. is provided for the incorporation of modified source code. From inspection of the original source the user identifies the section to be modified. This is not a task to be undertaken lightly. Although the source is generally well commented, it is voluminous. It may be necessary to refer to the PAFEC-FE system manual published by PAFEC Ltd.

Having copied and subsequently modified the subroutine or subroutines involved into a separate disk file, the contents of that disk file are compiled and loaded by the local operating system in preference to existing subroutines of the same name within PAFEC-FE by means of the USE. option. Although implementation of the USE. option may vary with different operating systems, in general everything should run smoothly if 'USE.' is followed by the name of the disk file containing the modified subroutines.

As an example, if the disk file 'MODS' contains the modified subroutines then the following control section would cause them to be incorporated into the PAFEC-FE system.

```
CONTROL
USE.MODS
CONTROL.END
```

3.5.5   Local options

In addition to the facilities provided by the standard PAFEC-FE program it is open to system programmers to add extra facilities or modify existing ones. Given access to the FORTRAN source code it is possible to tailor the program to meet a specific requirement. For example, the user may wish to obtain intermediate results from the stress calculations or to add a new finite element to the system.

Usually it is through local control options that differences between the standard PAFEC-FE program and its local implementation will be noticed. Some installations may find it necessary to modify the standard options in order to facilitate implementation of the PAFEC-FE program on non-standard operating systems. Such modifications will generally apply to the save and re-start facilities which are inevitably operating system dependent. Options which are liable to have been modified locally are marked with an asterisk (*) in the PAFEC-FE Data Preparation Manual and are shown in Appendix A.

Some installations may supply additional facilities through local control options. For example, there may be options available relating to size, quality and colour of PAFEC-FE drawings.

3.6 Comments

The PAFEC-FE program allows the user to annotate data sets, a practice which is to be encouraged. Any data record of which the first non-blank character is a 'C' followed by one or more blanks is 'echoed' by the program and no further action is taken.

For example, records such as the following are effectively ignored by the PAFEC-FE program and are only inserted by the user for documentation purposes.

C
C The following records define the structural nodes.
C

3.7 The end of data marker

The complete set of data must be terminated with a special end of data marker namely, a record:

END.OF.DATA

Once again note that the abbreviation END. (but not END) is permitted.

3.8 Conclusion

The aim of this chapter has been to present a broad outline of the PAFEC-FE program with indications of how it is to be used. The general principles involved in writing data modules have been examined. The number of PAFEC-FE modules is such that it would be impossible to look at each one in turn and so the reader is referred to the PAFEC-FE Data Preparation Manual for further details.

Reference

PAFEC-FE Data Preparation Manual, PAFEC Ltd., Strelley Hall, Strelley, Nottingham NG8 6PE.

4 The finite element mesh

4.1 Introduction

The specification of the finite element model has all the potential for presenting unsurmountable practical difficulties. Essentially the problem is the specification of the model in terms of key points, or nodes in order to define the shape of the structure under analysis.

It will be appreciated that as the degree of approximation increases the number of nodes to be specified could be measured in hundreds if not thousands. The problem is compounded if as a result of analysis it becomes apparent that a more accurate, or refined, mesh is required. Clearly it would be unacceptable to leave such a task wholly to the finite element program user. For this reason PAFEC-FE in common with other systems incorporates many labour saving devices. Although this chapter will be initially concerned with the methods for generating nodes on a piecemeal basis it will lead on to techniques for automating the mesh generation process. With a little experience the user will be in a position to generate quite complex structures easily and economically.

In this chapter we will disregard units of measurement and deal only in pure numbers. As discussed in section 3.2 this is the form in which data is presented to PAFEC-FE.

Although we will, as indicated in section 2.2, be concerned with the construction of a PAFEC-FE data set, we mention once again that the facilities of PIGS, the Pafec Interactive Graphics System offer even further assistance to the user.

4.2 Construction of a simple structure

As an illustrative example we will consider a plate structure to be modelled by two rectangular finite elements.

The model is shown in Figure 4.1. The two finite elements will be identified by the positions of their corner nodes, which are numbered 1 to 6 as shown. The allocation of a node number to a particular node is entirely the choice of the user. Any other set of positive numbers, and not necessarily consecutive numbers could have been chosen. The only requirement is different numbers for different nodes.

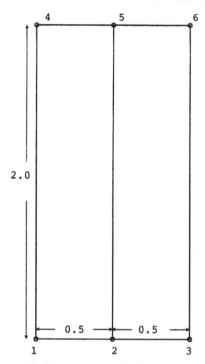

Figure 4.1 Plate structure showing node numbering.

In the example we will use Cartesian coordinates to specify the exact positions of the nodes. As with the node numbering, the positioning of the axes and by implication the position of the origin is left to the user.

4.2.1 The NODES module

Nodal positions are specified to the PAFEC-FE program using the NODES module. We have already seen this in section 3.4.2. Information concerning the node numbers, the type of coordinates and the actual coordinate values is supplied under the NODE.NUMBER, AXIS.NUMBER and X, Y and Z headers.

In our case since we are in two dimensions we will not need the Z header. We may also omit the AXIS.NUMBER header since the default value is 1, signifying Cartesian coordinates.

If we choose the x-axis to run in the direction of node 1 to node 3 and the y-axis to run in the direction from node 1 to node 4 and if we choose the origin to be at node 1, a NODES module defining a model of the structure could be

NODES
NODE.NUMBER	X	Y
1	0.0	0.0
2	0.5	0.0
3	1.0	0.0
4	0.0	2.0
5	0.5	2.0
6	1.0	2.0

Note that the positioning of the axes is implicit in our definition of the nodes.

A further simplification in the NODES module is possible using the automatic consecutive numbering feature of NODE.NUMBER. By default the numbering starts at 1 and increments by 1 with each new data line. So, the original nodes module could be written

NODES
X	Y
0.0	0.0
0.5	0.0
1.0	0.0
0.0	2.0
0.5	2.0
1.0	2.0

4.2.2 The ELEMENTS module

Having defined the outline of the structure we can now turn our attention to the specification of the finite elements to be used in the analysis. Bearing in mind the assumptions inherent in our model and the type of analysis required, the element type is chosen from the range described in the Data Preparation Manual. Subsequent chapters on the application of the finite element method will indicate how the choice is to be made.

The position of the individual elements within the structure is presented to the PAFEC-FE program by means of the ELEMENTS module. Node numbers are used to specify the position of each individual element. In this way the whole structure is defined.

At the simplest level, allowing the material of the structure to take the PAFEC-FE default option of mild steel we need only specify the element type and its position in the mesh, the relevant headers being ELEMENT.TYPE and TOPOLOGY.

As an illustrative example we will use the eight noded isoparametric quadrilateral curvilinear element, or element type 36210 as it is referred to by PAFEC-FE. Element type 36210 is specified by 8 nodes, as shown in Figure 4.2 reproduced here from the Data Preparation Manual. The numbers in Figure 4.2 are to be regarded as ordinal numbers, that is 1st node, 2nd

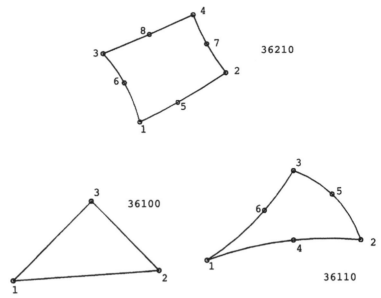

Figure 4.2   Isoparametric curvilinear elements for plane stress, plane strain and axisymmetric problems.

node, 3rd node and so on, and are used to pin-point the position, or topology of each element in the overall structure via the TOPOLOGY header. Referring to Figure 4.1, the 1st node of the left-hand element corresponds to what we have chosen as node 1, the 2nd node corresponds to what we have chosen to be node 2, the 3rd node to node 4 and the 4th node to node 5. For the right-hand element we have the 1st node is node 2, the 2nd is node 3, the 3rd is node 5 and the 4th is node 6.

Assembling this information we may write the ELEMENTS module modelling Figure 4.1 as

```
ELEMENTS
ELEMENT.TYPE  TOPOLOGY
   36210         1  2  4  5
   36210         2  3  5  6
```

In this example we do not specify 5th, 6th, 7th and 8th nodes. By default PAFEC-FE will automatically create nodes and allocate node numbers at the mid-points of the sides for these positions in the topology. This default option has the effect of creating two-dimensional elements with straight edges. To create an element with a curved edge the appropriate nodes between the relevant corners would be given specific coordinates by the user. PAFEC-FE would then create curved edges by constructing arcs through the two corner nodes and the intermediate nodes.

4.2.3 Irregular shapes

As indicated above the list of nodes under the TOPOLOGY header may be used to distort an element from the regular position. The program, however, may limit the distortions to reasonable proportions. A zero entry in the appropriate position is interpreted as a request for a node to be positioned at the mid-side, anything else is assumed to be the number of an existing node. For example, using element 36210 and assuming mild steel, the structure of Figure 4.3 might be modelled using

```
NODES
NODE.NUMBER   X     Y
    1             0.0   0.0
    2             0.5   0.0
    3             1.0   0.0
    4             0.0   2.0
    5             0.5   2.0
    6             1.0   2.0
   10            -0.1   1.0
   11             1.1   1.0
   12             0.75  2.1
ELEMENTS
ELEMENT.TYPE = 36210
TOPOLOGY
   1  2  4  5  0  10
   2  3  5  6  0  0  11  12
```

4.2.4 Material properties

A PROPERTIES header is provided in the ELEMENTS module so that element modelling materials other than mild steel may be incorporated. A number under the PROPERTIES header refers to an entry in either a PLATES.AND.SHELLS module, a MATERIAL module or a SPRINGS module depending upon the type of element chosen. That same number is itself an entry under the appropriate header of the module referred to. This method of linking (first mentioned in section 3.4) is quite a common feature of PAFEC-FE. The number to be used for linking purposes is chosen by the user and is quite unconnected with any other data item even though it may have the same value.

There are a number of standard material types in PAFEC-FE. For example, standard material number 4 is aluminium. To specify the previous mesh using aluminium would require the ELEMENTS module to refer to a PLATES.AND.SHELLS module as shown below

```
ELEMENTS
ELEMENT.TYPE = 36210
PROPERTIES = 3
```

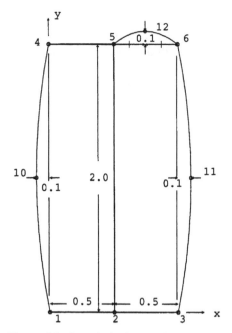

Figure 4.3 Irregularly shaped plate structure.

TOPOLOGY
  1 2 4 5
  2 3 5 6
PLATES.AND.SHELLS
PLATE.OR.SHELL.NUMBER   MATERIAL.NUMBER
   3   4

The linking facility allows different materials to be incorporated within the same model. For example, if the left-hand plate of Figure 4.1 was mild steel and the right aluminium we might have

ELEMENTS
ELEMENT.TYPE = 36210
PROPERTIES TOPOLOGY
  3 1 2 4 5
  4 2 3 5 6
PLATES.AND.SHELLS
PLATE.OR.SHELL.NUMBER   MATERIAL.NUMBER
   3   1
   4   4

Finally if we wished to incorporate non-standard materials, the MATERIAL module is available for the specification of particular values of Young's modulus, Poisson's ratio, mass density and other relevant factors.

For example, if a material having a Young's modulus of 25E9 and Poisson's ratio 0.25 replaced the mild steel of the left-hand plate, we would have a data set including modules of the following form

```
ELEMENTS
ELEMENT.TYPE = 36210
PROPERTIES TOPOLOGY
   3  1  2  4  5
   4  2  3  5  6
PLATES.AND.SHELLS
PLATE.OR.SHELL.NUMBER   MATERIAL.NUMBER
   3                         11
   4                          4
MATERIAL
MATERIAL.NUMBER  E      NU
   11                25E9  0.25
```

In this case we have chosen 11 as the link number. Although the choice is arbitrary, choosing a number between 1 and 10 is not recommended since there may be some confusion with one of the standard 10 material types in PAFEC-FE.

4.3 Modelling using polar coordinates

As an example we will consider the specification of a model having the shape of a quadrant as shown in Figure 4.4. We will assume that the quadrant is to be modelled by four equal straight sided triangular elements of type 36100. The element is shown in Figure 4.2. Clearly Cartesian coordinates could be used but polar coordinates are a more natural choice and will in this and similar cases simplify the specification of the nodes. To use polar coordinates we supply details via the AXIS.NUMBER header of the NODES module and the usual X, Y and Z headers. There are a number of axes sets available in PAFEC-FE each having an associated type number. For this example, we will use cylindrical polars, which in PAFEC-FE is referred to as type 2.

Although the header names X, Y and Z are more meaningful when applied to Cartesian coordinates, the same names have to be used for polars. In our example, using the type 2 cylindrical axis set, X refers to the distance along the axis of the cylinder, Y represents the radial measurement and Z represents the angular clockwise displacement in the plane of the radius. For our two-dimensional problem the distance along the axis of the cylinder will be zero.

Taking the origin at the centre of curvature we could specify the mesh with the following modules:

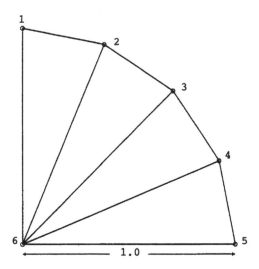

Figure 4.4 Quadrant shaped plate modelled by four elements.

```
NODES
AXIS.NUMBER = 2
NODE.NUMBER   X     Y     Z
   1          0.0   1.0   0.0
   2          0.0   1.0   22.5
   3          0.0   1.0   45
   4          0.0   1.0   67.5
   5          0.0   1.0   90
   6          0.0   0.0   0.0
ELEMENTS
ELEMENT.TYPE = 36100
TOPOLOGY
   6  2  1
   6  3  2
   6  4  3
   6  5  4
```

4.3.1 More complicated shapes

In practice it is often the case that shapes to be modelled have a combination of flat and curved edges and surfaces. Inevitably there may not be a single coordinate system which can be applied in a natural manner to the whole structure. Cartesians may be appropriate for the flat surfaces whilst polars may be more useful elsewhere. The AXES module is used for establishing differing coordinate systems in such a way that a new axis set is defined relative to some already established set.

As an example, consider the model outlined in Figure 4.5 having the node numbers as shown. Nodes 6–12 will be specified in Cartesian coordinates with an origin at node 6 whilst nodes 1–5 will be specified in polar coordinates with an origin at node 12.

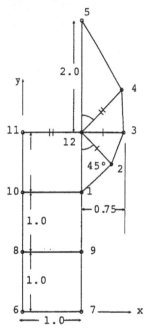

Figure 4.5 Plate structure.

In order to distinguish the two systems when specifying nodal coordinates we will use the AXIS header in the NODES module. An entry of 1, which is the default value will be used to refer to the standard Cartesian axis. The polar set, which is non-standard since the origin does not coincide with the origin of the Cartesian set will be designated by an entry of 20 under AXIS, the number 20 being a completely arbitrary choice. We will now show how our new axis set, which we have called axis set 20 is defined to the program.

As in section 4.3 we will use the cylindrical polar system provided by PAFEC and known to the program as type 2. To position this system we will use the AXES module. The entries under AXIS.NO, TYPE and NODE are 20, 2 and 12, respectively, indicating that we have designated a new axis set number 20, of type 2 with an origin at node 12. The entry under the RELAXISNO header is used to indicate the axis set relative to which the new set is defined. The necessary rotations about that relative set are entered under the ANG1, ANG2 and ANG3 headers. In our case the RELAXISNO entry is 1, to indicate the original Cartesian axis.

We now consider how those rotations are represented to the AXES module. Rotations are specified to the AXES module by means of the ANG1,

ANG2 and ANG3 headers. The new axis set is formed by a rotation of the original set by a rotation of ANG1 about the z-axis, followed by a rotation of ANG2 about the resulting y-axis followed by ANG3 about the latest x-axis. The names x, y and z refer to the relative axis and unfortunately are not the most meaningful when dealing with polars. We saw in section 4.3 that the polar system of type 2 is such that the axis of the cylinder lies along the Cartesian x-axis and the radial direction lies along the y-axis. In our example a rotation of 90° in the clockwise sense about the y-axis would rotate the cylinder so that its radial plane would lie in the original Cartesian xy-plane and at the same time point the axis into the paper and so the values for ANG1, ANG2 and ANG3 are 0, 90 and 0, respectively.

To summarize, using element types 36210 and 36110 the mesh shown in Figure 4.5 could be achieved by the following modules:

NODES

NODE.NUMBER	AXIS	X	Y	Z
6	1	0.0	0.0	
7	1	1.0	0.0	
8	1	0.0	1.0	
9	1	1.0	1.0	
10	1	0.0	2.0	
11	1	0.0	3.0	
12	1	1.0	3.0	
1	20	0.0	1.0	180
2	20	0.0	0.75	135
3	20	0.0	0.75	90
4	20	0.0	1.0	45
5	20	0.0	2.0	0

AXES

AXIS.NO	RELAXISNO	TYPE	NODE.NO	ANG1	ANG2	ANG3
20	1	2	12	0	90	0

ELEMENTS

ELEMENT.TYPE	TOPOLOGY			
36210	6	7	8	9
36210	8	9	10	1
36210	10	1	11	12
36110	12	1	2	
36110	12	2	3	
36110	12	3	4	
36110	12	4	5	

4.4 Automatic node specification

Although it is always possible to specify nodes on an individual basis there are occasions when it is more efficient to take advantage of the automatic node generation features of PAFEC-FE.

4.4.1 Line nodes

The automatic generation of equally spaced nodes along a line may be achieved through use of the LINE.NODES module.

The relevant header is LIST.OF.NODES.ON.LINE but as we have already noted, we may take advantage of the abbreviation facility and refer to this as LIST.

As an example, the nodes of Figure 4.6 could be generated by the following NODES and LINE.NODES modules. The LINE.NODES module equally spaces nodes 18, 19, 30 and 40 on the straight line joining nodes 16 and 20. Assuming node 16 to have Cartesian coordinates (1.0, 0.5) we have

```
NODES
NODE.NUMBER   X    Y
    16       1.0  0.5
    20       3.0  1.0
LINE.NODES
LIST
    16  18  19  30  40  20
```

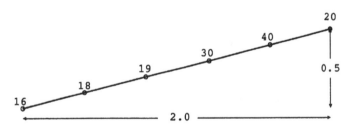

Figure 4.6 Example to illustrate the use of LINE.NODES.

4.4.2 Arc nodes

A similar effect is available for circular arcs. In this case the relevant module is ARC.NODES with the header LIST. The chosen circular arc is such that it passes through the first and last of the nodes in the list and is a least squares fit to nodes already defined. Apart from the first and the last, there will have to be at least one other node which will have to exist already in order to specify the arc. Where necessary, adjustments are made to existing nodes to force them onto the arc.

As an example, consider Figure 4.7 for which a combination of LINE.NODES and ARC.NODES will be used. Nodes 2 and 11 are used to define the circular arcs.

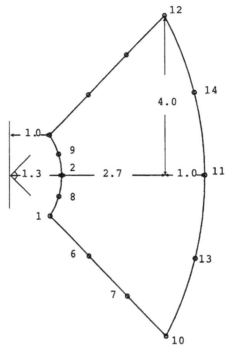

Figure 4.7 Example to illustrate the use of ARC.NODES.

Assuming Cartesian coordinates with an origin at the apex of the sector we have

```
NODES
NODE.NUMBER    X     Y
      3       1.0   1.0
      2       1.3   0.0
      1       1.0  -1.0
     12       4.0   4.0
     10       4.0  -4.0
     11       5.0   0.0
LINE.NODES
LIST
   3   4   5   12
   1   6   7   10
ARC.NODES
LIST
   1    8    2    9    3
  10   13   11   14   12
```

4.5 Mesh verification

As with all computer programs it is all too easy to make a mistake in the input data and so in effect solve or attempt to solve a different problem. Even though a program appears to work there is an obvious danger in unquestioningly accepting whatever results appear. In many cases incorrect or unacceptable results can be traced back to defects in the mesh. Mesh generation is not only one of the most complex tasks in finite element analysis, it can also be the most critical. Even with the economies outlined in this chapter it can still be a laborious task and as a result is prone to error.

There are two approaches to the reduction and elimination of such error, the first we examine is automatically provided by the PAFEC-FE program while the second is under the control of the user.

4.5.1 Element distortion

The accuracy of finite element analysis may be adversely affected by the use of unreasonably proportioned elements. The underlying mathematics of the element may not be able to handle distortions accurately. Although it is impossible to lay down hard and fast rules there are guidelines based on experience built into the PAFEC-FE program.

For all elements, the program calculates the ratio of the longest side to the shortest side. If that ratio is ever greater than 15:1 an error message appears and execution is terminated. If the ratio is less than 15:1 but greater than 5:1 a warning is given.

In a similar manner the program calculates the angles between adjacent sides of elements. If an angle is ever less than 15° or greater than 165° an error message appears. If the angle falls outside the range 45° to 135° but is still within the 15° to 165° range a warning is given.

Although warning as opposed to error messages do not in themselves signal a premature termination of the program they should not be ignored. They may signal inaccuracies in the final results. Ideally the mesh should be modified to avoid warning messages.

4.5.2 Node connectivity

In constructing a mesh it is important to be aware of the consequences of unconnected nodes. For example a corner node or mid-side node which does not coincide with a node of any other element will be treated by PAFEC-FE as a boundary node. Such nodes within the constraints of the problem are free to move without any connectivity conditions. Whilst this is perfectly acceptable for nodes which form the natural boundary it is an unrealistic model for nodes which are interior to the structure and which would normally be expected to remain in contact with the original surrounding structure. This connectivity principle which was used in constructing the linear equations of the finite element example of Chapter 1 is used in the more general course. It follows that a mesh of the type shown in Figure 4.8 of eight noded elements should be avoided since the mid-side nodes of the smaller elements which are initially in contact with the larger element will not be constrained to remain so.

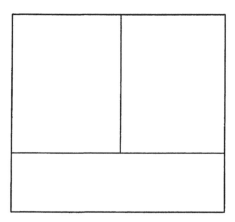

Figure 4.8 Example to illustrate lack of connectivity.

4.5.3 Model drawings

Although there are mesh verification procedures within PAFEC-FE it is still all too easy for an incorrectly modelled structure to slip through the net. In the last analysis it is the user who must decide. To produce a drawing is often the easiest and most effective means of verifying a generated mesh.

In PAFEC-FE, model drawing is controlled by data supplied to the IN.DRAW module. The headers available include TYPE.NUMBER for specifying the type of drawing (solid line boundary, broken line interior or whatever) and INFORMATION.NUMBER for specifying the information to be drawn (node numbers and element numbers being the most often required).

As an example the following IN.DRAW module would cause a solid line drawing of the model showing node numbers to be produced.

```
IN.DRAW
TYPE.NUMBER    INFORMATION.NUMBER
   2                  3
```

In general the PAFEC-FE program will produce a disk file containing drawing information. It is then for the local operating system to take this file and produce a drawing on whatever hardware is available.

4.6 Automatic mesh generation

So far we have used the facilities of PAFEC-FE to generate the finite elements one by one. Although this is reasonable for small structures it is clearly impractical for larger structures where hundreds if not thousands of elements may be involved.

The PAFBLOCKS module in conjunction with the MESH module provides a very powerful element generation facility. In a manner similar to the ELEMENTS module the PAFBLOCKS module by means of the TOPOLOGY header specifies the region, or block which is to be meshed. The type of element to be used is specified via the ELEMENT.TYPE header. The actual details of the mesh, the precise number and sizes of the subdivisions within the block is specified in a MESH module which is linked to the PAFBLOCKS module in the usual PAFEC-FE manner, described earlier in section 4.2.4.

As an example we consider a rectangular region having corner nodes as shown in Figure 4.9. Our aim will be to cover the region with elements of type 36210 by means of a 4 by 3 mesh as shown. The numbers 4 and 3 will appear in the MESH module under the SPACING.LIST header. They will be linked via headers N1 and N2 in the PAFBLOCKS module to entries under a REFERENCE header in the MESH module. As before the numbers used for linking are chosen arbitrarily. N1 refers to the side of the block formed by the 1st and 2nd nodes of the topology, in this example nodes 10 and 12, whilst N2 refers to the side formed by the 1st and 3rd nodes, in our case the actual node numbers are 10 and 13.

The mesh, as described might be achieved by the following modules. We assume Cartesian axes with node 10 at (1.0, 1.0).

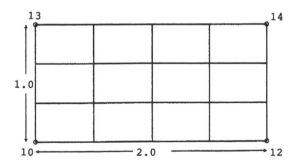

Figure 4.9 Rectangular region having a regular mesh.

```
NODES
NODE.NUMBER   X     Y
    10             1.0   1.0
    12             3.0   1.0
    13             1.0   2.0
    14             3.0   2.0
PAFBLOCKS
ELEMENT.TYPE  N1   N2   TOPOLOGY
    36210        2    3     10  12  13  14
MESH
REFERENCE   SPACING.LIST
    2            4
    3            3
```

The numbers 2 and 3 appearing in both the PAFBLOCKS and the MESH modules are for linking purposes and as such have been chosen arbitrarily.

The SPACING.LIST header may be used to produce an irregular mesh. If anything other than a single entry appears then these entries are interpreted to be the required ratios of the lengths of the elements.

For example, returning once more to our original block the mesh of Figure 4.10 could be achieved by means of the following modules:

```
PAFBLOCKS
ELEMENT.TYPE   N1   N2   TOPOLOGY
    36210         2    3    10  12  13  14
MESH
REFERENCE   SPACING.LIST
    2    1   2   1
    3    3
```

It can be seen from the examples that once a PAFBLOCK module is defined, major changes in the mesh can be effected by very simple changes under the SPACING.LIST header of the MESH module. This powerful feature of PAFEC-FE facilitates experiments with different mesh sizes, an essential practice in successful finite element analysis.

4.6.1 More generally shaped blocks

Although for illustrative purposes we have restricted our attention to two-dimensional blocks it is possible to mesh other shapes including two-dimensional triangular regions and three-dimensional block and wedge shapes. The type of region is specified via the TYPE header. In all the examples so far we have been implicitly assuming TYPE = 1, the default option.

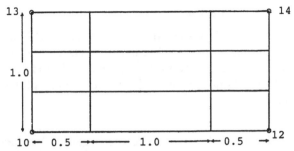

Figure 4.10   Rectangular region having an irregular mesh.

A block may be distorted and meshed so that individual elements reflect the overall distortion. Block distortion may be achieved through information supplied via the TOPOLOGY header in exactly the same way as outlined in section 4.2.3 for the ELEMENTS module. Figure 4.11 shows such a meshing based on the model of Figure 4.3. Given an appropriate nodes module the mesh could be achieved using the following modules:

```
PAFBLOCKS
ELEMENT.TYPE = 36210
N1  N2  TOPOLOGY
1    2    1  2  4  5  0  10
1    2    2  3  5  6  0  0  11  12
MESH
REFERENCE   SPACING
   1  3
   2  4
```

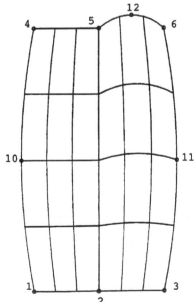

Figure 4.11   Mesh of a more generally shaped block.

4.6.2   Block material properties

A PROPERTIES header is provided in the PAFBLOCKS module for the specification of material properties. The use of this header in a PAFBLOCKS module is very similar to the use in the ELEMENTS module as outlined in section 4.2.4. The use of this header will feature in many of the applications discussed in later chapters.

4.7   Advanced features of PAFEC-FE mesh generation facilities

4.7.1   Modelling structures with holes or cutout sections

As an example we consider the uniform meshing of a plate having the shape shown in Figure 4.12. Although it would be possible to consider the use of separate PAFBLOCKS and separate ELEMENTS, we illustrate an alternative approach. In this and more complicated cases it is easier to mesh the structure as if it were a complete block and then remove unwanted elements. We do this in PAFEC-FE by means of the REFERENCE.IN.PAFBLOCKS module. The REFERENCE.IN.PAFBLOCKS module has headers C1 and C2 which act as pointers or coordinates to elements in an already established PAFBLOCK.

Assuming elements of type 36210, default material properties and that nodes 10, 20, 30 and 40 had been defined in a NODES module, we could generate the appropriately meshed rectangular block with those nodes as corners by the following modules:

```
PAFBLOCKS
ELEMENT.TYPE = 36210
BLOCK.NUMBER  N1  N2  TOPOLOGY
    1          1   2   10  20  30  40
MESH
REFERENCE  SPACING.LIST
    1          8
    2          4
```

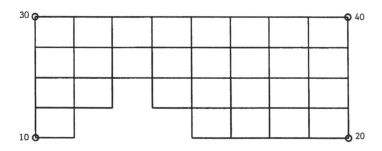

Figure 4.12 Example of element removal.

We now proceed to remove elements using the REFERENCE.IN. PAFBLOCKS module. The BLOCK.NUMBER header has been used in PAFBLOCKS to identify the PAFBLOCK to which we will refer. It is quite likely that in some problems we may have more than one PAFBLOCK.

The four elements in question are identified by the coordinates (2,1), (3,1), (4,1) and (3,2) in the N1 and N2 directions. These coordinates appear under the C1 and C2 headers of the REFERENCE.IN.PAFBLOCKS module. An entry of 1 under the REMOVE header signifies removal. Since the removal entries and the block number entries are constant throughout it is more economical to use the constant property feature and write the module for removal of the required elements as

```
REFERENCE.IN.PAFBLOCKS
BLOCK.NUMBER = 1
REMOVE = 1
   C1  C2
   2   1
   3   1
   4   1
   3   2
```

4.7.2 A mesh within a mesh

We have already referred to the necessity in finite element analysis of successively refining the mesh until a position of numerical stability in the results of the analysis is achieved. Wherever the structure to be modelled is reasonably simple in outline and where the external forces are applied fairly evenly, the use of one or more PAFBLOCKS module with refinement controlled by the MESH module should suffice. In such cases the results of the analysis would be expected to vary uniformly across the whole structure. On the other hand if the structure is a complicated one with perhaps intense local activity the results of the analysis may be expected to vary considerably over relatively small intervals. Although in theory it would be possible successively to refine the whole structure this might make unacceptable demands on computer time and memory. In practice local areas of intense activity are refined to a higher degree. The REFERENCE.IN.PAFBLOCKS module may be used for identifying elements within an already PAFBLOCKS generated mesh with a view to replacement by a finer element mesh. We consider the model and the required mesh as shown in Figure 4.13. We will assume that the results are expected to be particularly sensitive in the region of the denser mesh. It may be that external forces are particularly active in this region or that engineering intuition suggests the occurrence of major stresses.

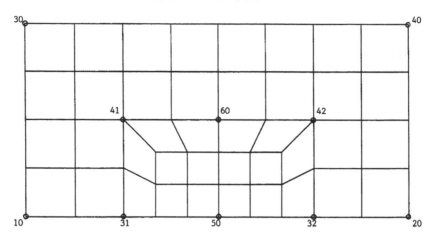

Figure 4.13 Mesh within a mesh.

We note that some ingenuity may be required in constructing a localized mesh refinement. It is important to bear in mind the connectivity principles discussed in section 4.5.2 and so avoid the generation of spurious boundary nodes.

To achieve the desired mesh, we first cover the whole block with a rectangular mesh and then remove the elements from the region we wish to refine. We then use another PAFBLOCK to define the mesh in the region we have just cleared. In order to define the second mesh we will require node numbers at the relevant corners. These may be defined using the headers POSITION and NODE in the REFERENCE.IN.PAFBLOCKS module. POSITION refers to the position in the topology of the element specified by the C1 and C2 values. NODE is the node number we wish to assign. For example, in Figure 4.13 we wish to name node 31 as being at topology position 1 in the element having C1, C2 values (3,1). Equally well we could regard node 31 as being at topology position 2 in element (2,1). For a node such as 41 it can be seen that there are four possible positions, namely position 4 in element (2,2), 3 in (3,2), 2 in (2,3) and 1 in (3,3).

Given that we wish to mesh the refined region using two type 5 PAFBLOCKs we could specify the following data modules. We will assume element type 36210, default material of mild steel and that the node numbers 10, 20, 30 and 40 have been defined in a suitable NODES module.

To specify the overall mesh we have

PAFBLOCKS
ELEMENT.TYPE = 36210

```
BLOCK.NUMBER  N1  N2   TOPOLOGY
   1           1   2    10  20  30  40
MESH
REFERENCE   SPACING.LIST
   1  8
   2  4
```

To remove specific elements and at the same time allocate the node numbers 31, 50, 32, 41, 60 and 42, we have

```
REFERENCE.IN.PAFBLOCKS
BLOCK.NUMBER = 1
REMOVE = 1
C1   C2   POSITION   NODE
3    1    1          31
4    1    2          50
5    1
6    1    2          32
3    2    3          41
4    2    4          60
5    2
6    2    4          42
```

And finally noting from the Data Preparation Manual that a type 5 PAFBLOCK requires an N3 specification in addition to the usual N1 and N2, we might establish the inner mesh using the following PAFBLOCKS and MESH module:

```
PAFBLOCKS
ELEMENT.TYPE = 36210
TYPE = 5
BLOCK.NUMBER  N1  N2  N3   TOPOLOGY
   2           3   4   5    50  32  60  42
   3           3   4   5    50  31  60  41
MESH
REFERENCE   SPACING.LIST
   3           2
   4           2
   5           1
```

In passing it is worth pointing out that the two PAFBLOCKS modules above could be combined into a single PAFBLOCK, namely

```
PAFBLOCKS
ELEMENT.TYPE = 36210
BLOCK.NUMBER  TYPE  N1  N2  N3   TOPOLOGY
   1           1    1   2   0    10  20  30  40
   2           5    3   4   5    50  32  60  42
   3           5    3   4   5    50  31  60  41
```

A similar comment applies to the MESH modules. In fact since three of the values under the spacing list are identical the link numbers 2, 3 and 4 under the N1, N2 and REFERENCE headers could be replaced by a single number and so reduce the entries in the combined MESH module from five to three.

4.8 Conclusion

We have now completed our survey of the facilities available in PAFEC-FE for mesh generation. It is not the easiest of tasks, yet it is crucial to successful analysis. In the next chapter we will conclude the outline of the task of data preparation by indicating the modules to be used in the representation of external loads and pressures and other boundary conditions. In general these other modules are constructed along similar lines to those we have seen in this chapter. If anything, the constructions are simpler.

5 General problem solving strategy

5.1 Identification of the engineering problem

Although it may sound trivial and may appear self-evident, the importance of understanding the question, whatever the subject, cannot be overestimated. To understand the question fully is to go a considerable distance towards finding a solution. A well formed in-depth formulation of any problem will inevitably contain the seeds of the solution.

In formulating an engineering problem the important factors are the structure, the environment of the structure, the model of the structure, the analysis and the results required from the analysis. All the factors need to be carefully considered before finite element analysis or indeed any other type of analysis is deemed appropriate.

At a general level the structure has to be considered from the point of view of the geometry, boundary conditions and the material composition. At a local level restrictions on free movement, connections, joints and other special structural features have to be taken into account. Consideration of the environment of the structure will involve the specification of all external forces, pressures, loadings and supports and any required displacements and contact conditions. The availability and reliability of data will have to be assessed.

A decision has to be made as to the type of analysis required. There are standard analyses available such as linear and non-linear static displacement and stress analysis, thermal conduction and thermal stress analysis, and modal frequency and response analysis. In a minority of cases a manageable analytical solution is available but for the rest, the overwhelming majority, an approximation such as the finite element method is required. A non-standard analysis may require either a hybrid approach or an innovative approach. At all times the ultimate requirement, namely the nature of the required results, their scope and required accuracy will determine and influence the analysis.

5.2 Idealization as a model

In essence, the preceding section has outlined the steps to be taken in formulating an abstraction or model of the engineering problem. A model is essentially a copy in a simplified form of the real thing. A typical example is a model aeroplane, which may or may not fly. The essential quality of this and other models is that it presents an opportunity on which to perform experiments which might otherwise be either impossible or impractical. The underlying assumption is that the results from working with a model may be related to the original structure. This of course depends on the model faithfully reproducing, albeit in its own way, the characteristics of the original. In finite element analysis we are not concerned with physical models but rather conceptual models which may be represented by a collection of drawings and associated data. Inevitably in the transfer from reality to a drawing on paper with a list of data, a degree of authenticity is lost. Successful analysis will depend not only on the accuracy of the mathematics but on the accuracy of the modelling process. Modelling techniques are a study in themselves and offer enormous scope for individual ingenuity. Although subsequent chapters will show how individual problems have been modelled and subsequently analysed we cannot pretend that there are hard and fast rules to follow. We warn of the all too common mistake of regarding the model as the reality. In the face of unacceptable results it is all too tempting to take the softer option of tinkering with the model rather than questioning the model itself.

5.3   Realization of the model in PAFEC-FE

In formulating the model the separate and distinct characteristics of the problem will emerge. These characteristics are then translated into PAFEC terms. In general the separate measurable aspects of the model are represented to PAFEC-FE by data modules. A typical example is the NODES module which outlines the overall shape. The type of analysis chosen by PAFEC-FE is usually automatically determined by the type of modules present in the data, although occasionally extra guidance from options placed in a CONTROL may be required.

5.4   The structure

The specification of the geometry of the model and subsequent meshing has already been covered in chapter 4. In effect there are no limits to the shape and size of a structure which may be realized in PAFEC-FE.

The modules available for specifying the structure are:

- AXES: Establishing of coordinate systems
- NODES: Node specification
- LINE.NODES: Specification of a line of nodes

- ARC.NODES: Specification of a circular arc of nodes

The modules available for specifying the elements and the mesh are:

- ELEMENTS: Type and location of individual elements
- PAFBLOCKS: Element specification across a region
- MESH: Referenced by PAFBLOCKS for mesh specification
- REFERENCE.IN.PAFBLOCKS: Used for more sophisticated meshing

The modules required for specifying particular element characteristics and material properties are:

- MATERIAL: Material properties of specific elements
- BEAMS: Beam element specifications
- PLATES.AND.SHELLS: Plate and shell element specifications
- SPRINGS: Spring element specifications
- LAMINATES: Referred to by ELEMENTS module in the case of laminated materials
- ORTHOTROPIC.MATERIAL: Referred to by LAMINATES module
- FAILURE.CRITERIA: Definition of failure criteria for specific materials
- MASSES: Referred to by ELEMENTS module for defining element masses
- VARIABLE.MATERIAL: Specification of material properties varying with temperature or potential

5.5 The analysis

We look at the types of analysis available and the important relevant modules and control options. Normally a selection of modules and control options would be made.

5.6 Static analysis

5.6.1 Static displacement and stress analysis

A static analysis determines how the structure deflects and the stresses developed as static loads are applied. The modules available for describing static loads are:

- LOADS: Point loads at specific nodes
- MEMBER.LOADS: Distributed loads to beam elements
- PRESSURE: Pressure at specific nodes
- GRAVITY: Gravity and inertia loads
- CENTRIFUGAL: Centrifugal loads

- TEMPERATURE: Thermal loading for subsequent thermal stress calculation
- SURFACE.FOR.PRESSURE: Pressures over given surfaces
- FACE.LOADING: Pressures over specific element faces

The modules available for specifying required displacements, restraints and contact conditions are:

- DISPLACEMENTS.PRESCRIBED: Prescribed displacement at specific nodes
- RESTRAINTS: Restraint conditions at specific nodes
- ENCASTRE: Structure nodes which are fully restrained
- GENERALISED.CONSTRAINTS: Specification of rigid links
- HINGES.AND.SLIDES: Freedom of movement between connected nodes
- REPEATED FREEDOMS: Alternative form of HINGES.AND.SLIDES module
- GAPS: Specification of structural nodes which are allowed to separate with or without friction
- BALANCED.LOAD.CASES: Loading for zero reaction
- LOCAL.DIRECTIONS: Restraint conditions specified in local axes
- POLAR.DIRECTIONS: Specification of radial and tangential degrees of freedom
- RIGID.LINKS: Nodes which are to be rigidly linked
- SIMPLE.SUPPORTS: Nodes which are to be simply supported
- CRACK.TIP: Specification of a structural crack
- REMOVE.CONSTRAINTS.AND.LOADINGS: Detail of individual conditions to be removed following a more general specification; the use of this module may simplify data preparation

5.6.2   Non-linear static analysis

The types of non-linear analysis offered by PAFEC-FE include plasticity, creep, buckling and large displacement analysis. Precise definitions will be given in the next chapter. The analyses required are specified through the CONTROL options, the options available being PLASTIC, CREEP, BUCKLING and LARGE.DISPLACEMENT. Non-linear analysis simulates non-linearity by repeated linear approximations. This type of analysis may therefore be time consuming.

The relevant modules are:

- PLASTIC.MATERIAL: Defines the elasto-plastic material stress-strain and the strain hardening rule
- YIELDING.ELEMENTS: Specification of the elements which are allowed to yield and their associated plastic material properties

- INCREMENTAL: Specification of the manner in which load increments are to be applied in a plasticity or large displacement analysis; specifies the time step for an analysis
- CONVERGENCE: Convergence criteria for non-linear analysis
- CREEP.LAW: Definition of the creep law for the elements allowed to yield
- ITERATION: User control over number of iterations per time step in creep and plasticity analyses

The PAFEC-FE program provides a save and restart facility enabling an analysis to be completed over more than one execution of the program. This facility is particularly useful for non-linear problems which may be time consuming. The relevant CONTROL options are STORE.INCREMENT, STOP.INCREMENT and REDUCTION.RESTART for saving and re-starting within the Phase 7 solution phase and INCREMENT.STORE, INCREMENT.STOP and INCREMENT.RESTART for similar saving and re-starting at the Phase 9 stress calculation phase.

5.7 Thermal analyses

The following modules are available for all types of thermal analysis:

- TEMPERATURE: Initial temperature field
- HEAT.TRANSFER: Prescribed constant or varying thermal inputs

5.7.1 Steady state analysis

The modules to be used to calculate the steady temperature distribution include:

- FLUX: prescribed thermal inputs

5.7.2 Transient analysis with prescribed initial field

- THERMAL.SHOCK: Time dependent temperature variations
- NODAL.FLUX.SHOCK: Time dependent heat flux variations
- TIMES.FOR.THERMAL.STRESS.CALCULATION: Time during a transient analysis at which thermal stress analysis is required
- UNSTEADY.THERMAL.TIMES: Time step to be used in a transient analysis
- TRANSPORT.PARAMETERS: May be used to model changes in thermal conductivity
- ITERATION: Used in analyses where material properties vary with temperature

5.7.3 Steady state followed by transient analysis

If both options CALC.STEADY.TEMPS and CALC.TRANS.TEMPS are included in the CONTROL section a transient analysis follows a steady state analysis.

5.7.4 Thermal stress analysis

A thermal stress analysis has to be carried out by two separate executions of the PAFEC program. The first program, which determines the temperature distribution, is followed by a program for stress analysis. The temperatures calculated in the first analysis are saved for later use by placing the option SAVE.TEMPS.TO.filename in the CONTROL section, filename being the name of a suitable disk file chosen by the user. In the second program structural elements replace the thermal elements and modules referring to thermal properties are replaced by modules specifying the restraints on the structure. In order to retrieve the already calculated temperatures, the option READ.TEMPS.FROM.filename is placed in the CONTROL section.

5.8 Dynamics and vibration analysis

5.8.1 Natural frequencies and mode shapes

The modules available to determine the natural frequencies and mode shapes of a structure are:

- MODES.AND.FREQUENCIES: Number of full mode shapes required
- RESPONSE: Type of response, with or without damping required

5.8.2 Sinusoidal response

The relevant modules to determine the response of the structure due to loads which vary sinusoidally with time include:

- FREQUENCIES.FOR.ANALYSIS: Frequencies at which calculations are to be made
- SINE.LOADING: Specification of the loading
- TABLE.OF.APPLIED.FORCES: Frequency at which the loading is applied
- DAMPING: Specification of mode damping conditions
- LINKS.FOR.DYNAMICS: Specification of between node dampers

5.8.3 Transient response

The relevant modules to obtain the response to a more general loading include:

- VELOCITIES.PRESCRIBED: Nodes at which a velocity is to be prescribed
- FORCING: Times at which forces are to be applied
- CHANGE.OF.MASS: Times at which masses change in a transient response analysis
- DEFINE.RESPONSE: Times at which prescribed motions are applied

5.9 Other analyses

There are a number of other analyses available in the PAFEC-FE program including lubrication, piezoelectric, acoustic, magnetic field and seismic response analyses. Although these are rather specialized areas the data is constructed in the usual way from the range of modules and control statements provided.

For very large problems, PAFEC-FE has a sub-structuring facility. This facility enables structures to be considered as a composite of separate structures, each of which may be analysed separately before being merged to produce results for the whole structure. This is a more efficient use of computing resources in that individual components may be modelled and analysed in depth without at the same time repeating calculations for parts of the structure which remain the same. Indeed the advent of so-called multiprocessors, (computers with more than one processor) has added impetus to the use of sub-structuring techniques. Such computers permit more than one task to be performed simultaneously. In future it is likely that as such machines become the standard, all structures to be analysed by the finite element method will be dissected into component sub-structures. The sub-structures will then be analysed simultaneously in order to make most efficient use of the computer. Even allowing for subsequent combination of results from the component sub-structures, the time taken for a full analysis will be roughly equal to the time taken to analyse the most complex sub-structure, since while this is going on, everything else is being analysed simultaneously.

5.10 Units

As already indicated in section 3.2 data are entered into PAFEC-FE as dimensionless quantities. PAFEC-FE is programed to produce results which are consistent in SI units. If the user prefers to use other units the relevant constants must be overwritten by information supplied under the appropriate module headers.

5.11   The results

In the absence of information to the contrary, the program normally outputs all the calculations to disk files. The user may then interrogate the files or simply direct a listing to the line printer. There are, however, a number of modules and control options enabling the user to customize the extent and nature of the results. This may be necessary as the results from a PAFEC-FE analysis can be voluminous.

The following modules relate to printed output:

- STRESS.ELEMENT: List of the elements to be stressed; in the absence of information from this module the program carries out the stress calculations for all the elements in the structure
- REACTIONS: Nodes at which the reaction is to be calculated; if the calculation is required at every node the CONTROL option REACTIONS may be used
- STRAIN.ENERGY.DENSITY: Elements for which the results of strain energy density calculations are required
- PROCESSING.FOR.STRESS.OUTPUT: User control over the presentation of stress calculations
- EXTERNAL.FORCES: Coordinate system in which forces are to be resolved and then listed
- RESULTS.FOR.BEAMS: Prescribes the information required from beam element calculations

In addition to previously mentioned modules for the control of output, the following are particularly relevant to dynamics analysis:

- SINUSOIDAL.OUTPUT: User control over output from sinusoidal response
- FULL.DYNAMICS.OUTPUT: Intermediate times during a transient response analysis at which displacements, stresses and displaced shape drawings are required
- BUCKLING.HARMONICS: List of harmonics at which results are required in a buckling analysis

The following modules generate graphical output:

- IN.DRAW: Graphical display of the model prior to analysis.
- OUT.DRAW: Graphical output of the results of the analysis from a global viewpoint
- SELECT.DRAW: May be used in conjunction with IN.DRAW and OUT.DRAW to restrict the number of elements drawn
- SECTION.FOR.PLOTTING: Used in conjunction with OUT.DRAW to produce cross-section graphs of stress contours
- GRAPH: More localized graphical output of the results

- PLOTS.OF.TEMPERATURE.FRONT: Graphs of temperature fronts
- DYNAMICS.GRAPH: Frequency and response graphs
- ADDITIONAL.GRAPH.INFORMATION: Temperature and time graphs

5.12 Has the problem been solved?

It is all too easy to yield to a sense of relief when all the syntactical errors in the data have been corrected and the program finally runs to completion. In such circumstances it is tempting but decidedly unwise and indeed potentially dangerous to accept the results without further question.

At the very least another analysis should be carried out with a more refined mesh. In theory the results of the analysis should converge as the density of the mesh increases. Although in practice the limitations of computer arithmetic may prevent total convergence it should be possible to detect a trend in the results towards a stable position as the number of elements increases. If this does not happen it is likely that the problem is numerically ill-conditioned which in turn leads to a questioning of the assumptions inherent in the model. Intuition suggests that the mathematics applied to an unstable model are themselves unstable. The boundary conditions, the choice of element type and the loadings should all be carefully reconsidered. A complete re-design of the model may be necessary. On the other hand a realistic model of a physically sound structure should be reflected in equally stable mathematics.

Even if a position of numerical stability is reached the results for individual quantities such as displacements, stresses and temperatures should be examined for any sign of irregularity. Unless there are sound engineering reasons to the contrary all such quantities should vary across the region in a smooth manner, and be free from sudden jumps and discontinuities. If there is any doubt a return to the modelling process must be made.

Finally, the user must exercise engineering judgement based on experience and on the experience of others to decide if the results make sense. In this context the wider experience of others will include the results published in the appropriate journals.

The users of the PAFEC-FE program have the assurance that they are using an internationally recognized product based on principles recommended by the National Association for Finite Element Methods and Standards (NAFEMS). However, it would be naïve to assume that everything that can be done has been done. The product is under constant review with new features being added on a regular basis. Errors and situations which might lead to errors are also regularly corrected although it is hoped that through better programming design methodologies and regular feedback from users that such malfunctions will eventually be removed. The company has always maintained an enlightened policy towards its users, recognizing the fact that their experiences are beneficial in enhancing the product. To this end PAFEC Ltd. listens to the suggestions made collectively through the independent users association. The company also maintains a finite element support desk which is able to offer advice to users having a valid maintenance agreement. The users association and the support desk are avenues to consider when local support and advice is to no avail.

6 PAFEC-FE finite elements and their constitutive equations

6.1 Introduction

We now look at some of the underlying mathematics of the more commonly used finite elements in PAFEC-FE. This will give an insight into the types of analysis that can be carried out and also the extent of the results that can be expected. The essence of this chapter, as indeed is the essence of finite element theory, is the stress–strain relationship in all of its various forms. It must be pointed out that in this chapter it is not possible to go into the detail of all the constitutive equations and thus some generalizations have to be accepted.

The constitutive (stress–strain) relationships as used in PAFEC-FE are similar to those used in classical engineering analysis. However, since a computer is to carry out the work, some relationships are made a little more complex and consequently a little more accurate. Two types of relationship are used: a linear elastic stress–strain relationship, where the stress is proportional to the strain as shown in Figure 6.1, and a non-linear stress–strain relationship, where the stress is not proportional to the strain as shown in Figure 6.2.

In general, stress $\sigma = K \times \varepsilon$, where ε is defined as strain and K is some constant depending upon the material being used. In linear elastic stress–strain, K is traditionally denoted by E, the Young's modulus, ν the Poisson's

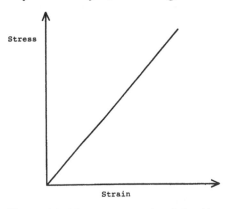

Figure 6.1 Linear stress–strain relationship.

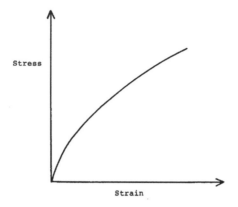

Figure 6.2 Non-linear stress–strain relationship.

ratio and, under uniaxial conditions is written as in equation (6.1):

$$E = \frac{\sigma}{\varepsilon} = \frac{\text{stress}}{\text{strain}} = \text{Young's modulus} \qquad (6.1)$$

In the PAFEC-FE system, the finite elements are developed in a series of stages that involve the constitutive relationships. The stages will be briefly described.

The basis of the finite element method is the matrix equation (6.2),

$$[K]\,[d] = [F] \qquad (6.2)$$

which is similar to the constitutive relationship, where $[F]$ is the vector of nodal loads, $[d]$ is the vector of nodal displacements and $[K]$ is the stiffness of the complete structure. We have already seen an example of this equation in section 1.2.2.

One of the advantages of the finite element method is that the sequence of operations involved in the method is independent of the type of finite element being used. This is important as any program for finite element analysis can be written to analyse a wide variety of problems, provided the appropriate finite elements with appropriate displacement functions and constitutive relationships are used.

The development of a finite element, which is described in almost all books on finite element theory (Zienkiewicz, 1989), proceeds through the following stages:

(1) The assumptions concerning the element geometry are listed.
(2) A displacement function is chosen to model the state of displacement at any point in the element. The displacement function is an example of a Courant trial function, described in the introductory chapter.

(3) The coefficients of the displacement function are now expressed in terms of the nodal displacements. It follows that the displacements at any point within the element are related to the nodal displacements.

(4) The strains ε at any point in the finite element are now related to the displacements at that point and hence to the nodal displacement equation.

(5) The internal stresses σ occurring in the element are now related to the strains ε through the elasticity matrix $[D]$ as shown in equation (6.3).

$$[D][B] = [\sigma], \text{ where } [B] \text{ is the matrix of strains} \qquad (6.3)$$

This is a generalization of the previously mentioned stress–strain equation.

(6) The nodal loads $[F]$ are related to the nodal displacements $[D]$ by using the constitutive relationships and thereby defining the required element stiffness matrix. If the nodal forces are defined in global coordinates rather than in element coordinates, then a transposition matrix between the two sets of coordinates will be required.

(7) Finally a stress displacement matrix is used to relate the internal stresses in the finite element to the nodal displacements. This enables the internal stresses to be evaluated.

It must be noted that only in very simple finite elements does the chosen displacement function correspond exactly to the correct displacement shape. In general the chosen displacement function will only approximate to the correct displaced shape and thus the resulting stresses and displacements will also only approximate to the correct value. For this reason the use of any finite element is invariably restricted to the use for which it was designed and within the distortion limits of its development. For example, a flat plate bending finite element may give incorrect stresses if it is used to model a thin cylindrical shell. Also a three-dimensional finite element may give incorrect stresses if certain dimensional constraints are ignored. The user of PAFEC-FE is therefore urged to be careful in the choice of element for a particular analysis, noting at all times, the restrictions and limitations imposed in PAFEC-FE.

Before going on to discuss the element types, it is necessary to say a little about the degrees of freedom at each node. Normally we use the three axes x, y and z, which are perpendicular to each other. Along each axis is a translation U_x, U_y and U_z, respectively, and a rotation θ_x, θ_y and θ_z, respectively, as shown in Figure 6.3. The axis system is right handed and the rotation, which is shown as a double headed arrow on the axis is in the right-handed screw direction looking from the origin of the axis.

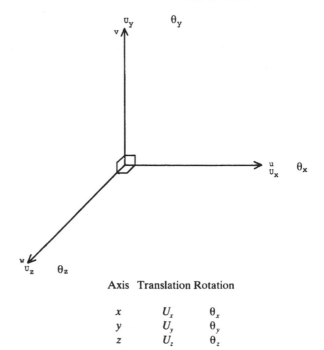

Axis Translation Rotation

Axis	Translation	Rotation
x	U_x	θ_x
y	U_y	θ_y
z	U_z	θ_z

Figure 6.3 Axis system.

6.2 Discussion of element type

To continue the discussion of section 2.5 we note that the different types of finite element may be categorized as follows:

(a) Beam elements which have two nodes, with only two degrees of freedom per node are considered as one-dimensional. Some beam elements may have up to six degrees of freedom per node.

(b) Membrane elements have three or more nodes which carry only in-plane loads. Each node has two degrees of freedom, U_x and U_y. These elements are considered two-dimensional elements.

(c) Three-dimensional elements have four or more nodes with three degrees of freedom, U_x, U_y and U_z per node.

(d) Plate bending elements have three or more nodes, with one transverse U_z and two rotational degrees of freedom, ϕ_x and ϕ_y per node.

(e) Shell elements have three or more nodes with up to three translational degrees of freedom U_x, U_y and U_z, and three rotational degrees of freedom ϕ_x, ϕ_y and ϕ_z per node.

(f) Thermal elements with two or more nodes per element, where the temperature is the only degree of freedom per node.

(g) Elements for dynamic and vibration problems.

(h) Hybrid elements involving any combination of the above types are also possible. PAFEC-FE also allows for the incorporation of user defined elements.

6.3 Constitutive equations

We now look at the constitutive equations involved in various types of elements and used for a range of analyses.

6.4 Beam elements

In PAFEC-FE, beam elements are classified as follows:

(a) Simple beam element (e.g. type 34000), with two nodes and six degrees of freedom at each node; shear deformation and rotary inertia are neglected
(b) Straight beam element (e.g. type 34100), with shear deformation and rotary inertia, but otherwise the same as the preceding element
(c) Simple beam element (e.g. types 34200 and 34500), with offset, but with no shear deformation or rotary inertia
(d) Straight beam element (e.g. types 34210, 34600 and 34700), with offset, shear deformation and rotary inertia
(e) Curved beam element (e.g. types 34300), with shear deformation and rotary inertia

How to determine which element to use in a given analysis is described in the following chapters.

For the simple beam element, it is assumed that the beam is prismatic, that is, the cross-sectional details do not vary along the length of the element. It is further assumed that the flexural and shear centres lie along the line joining the two end nodes. Because of the simplicity of the bending motions applied to the element, the bending stiffness matrix is exact and can be derived in a number of ways, for example using the moment area method. Axial forces only become significant when the in-plane forces are not small compared with the load required to buckle the structure.

For straight elements with shear deformation and rotary inertia, shear deformation due to a shearing force is included.

The offset beam element is identical to the simple beam element, except that it is used where the shear and flexural centre of the element, does not coincide with the element node.

The curved beam element has a constant curvature with the in-plane and out-of-plane motions being uncoupled and considered separately.

Let us consider a simple beam element with two end nodes. Each node having a lateral displacement U_y, and a rotation ϕ_z as shown in Figure 6.4. This gives four degrees of freedom per node and requires a four term function to describe the displacement. The displacement function is shown in equation (6.4), which also includes the term for the rotation.

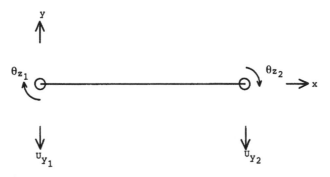

Figure 6.4 Simple beam element with two nodes and four degrees of freedom.

$$U_y = \alpha_1 + x\,\alpha_2 + x^2\alpha_3 + x^3\alpha_4$$

$$\phi_z = \frac{\partial v}{\partial x} = \alpha_2 + 2x\alpha_3 + 3x^2\alpha_4 \qquad (6.4)$$

The four term displacement function in equation (6.4) uses a polynomial expression that gives the most accurate results. In bending, the only significant strain component is the axial strain which varies linearly through the beam in the usual form of the equation as shown in equation (6.5).

$$\varepsilon_x = -y\left(\frac{\partial^2 v}{\partial x^2}\right) \qquad (6.5)$$

However, in beam analysis the constitutive relationship is determined through the curvature of equation (6.6).

$$\chi_z = \frac{\varepsilon_x}{y} = -\frac{\partial^2 v}{\partial x^2} = -2\alpha_3 - 6x\alpha_4 \qquad (6.6)$$

$$M_z = E\,I_z\chi_z = E\,I_z\,[-2\alpha_3 - 6x\alpha_4]$$

For loads applied at the nodes, the one-dimensional beam element has a stiffness matrix that is exact as derived through the use of the finite element method described so far. The beam is in simple bending and the three conditions of elasticity, compatibility of displacement and equilibrium of forces are satisfied throughout the element.

For elements that are not simple, the complete satisfaction of the above three conditions is unusual. The requirements of elasticity and compatibility are usually met, but the condition of equilibrium is often not met or not satisfied. This means that inaccuracies are introduced into the formulation of the constitutive equations and the resulting finite element can only be relied upon to give reasonable answers when used within its designed domain. For example flat plate bending finite elements would not be expected to give accurate stresses and displacements to a bending shell problem.

6.5 Membrane elements

Turn now to the membrane elements (e.g. types 36100 and 36200), which carry in-plane loads only and have two degrees of freedom, U_x and U_y per node. These elements are considered as two-dimensional elements with a constant thickness h. It is assumed that stresses do not vary through the thickness of the element. A typical element is shown in Figure 6.5.

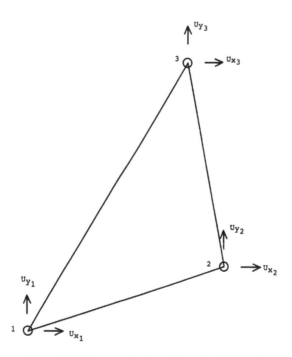

Figure 6.5 Three node, two-dimensional element.

The triangular element has three nodes and two degrees of freedom per node. Thus a six term polynomial displacement function is used in terms of x and y as shown in equation (6.7). Equation (6.8) gives the strains, which include not only the direct strains ε_x and ε_y, but also the shearing strain γ_{xy}.

$$U_x = \alpha_1 + x\alpha_2 + y\alpha_3$$
$$U_y = \alpha_4 + x\alpha_5 + y\alpha_6 \tag{6.7}$$

$$\varepsilon_x \quad = \quad \frac{\partial U_x}{\partial x} = \alpha_2$$

$$\varepsilon_y \quad = \quad \alpha_6 \tag{6.8}$$

$$\gamma_{xy} \quad = \quad \frac{\partial U_x}{\partial y} + \frac{\partial U_y}{\partial x} = \alpha_3 + \alpha_5$$

In the case of the beam element there is only one term in the elasticity matrix, namely the Young's modulus E. For membrane elements the constitutive matrix will include the Poisson's ratio term and will have different values depending upon whether 'plane strain' or 'plane stress' is being investigated. Plane strain assumes that there is no strain perpendicular to the plane, while plane stress assumes that there is no stress normal to surface; the bending stresses are also negligible. The elasticity matrix for both cases is shown in equations (6.9) and (6.10).

$$[D] = \frac{E}{1 - v^2} \begin{bmatrix} 1 & v & 0 \\ v & 1 & 0 \\ 0 & 0 & \frac{1}{2}(1 - v) \end{bmatrix} \tag{6.9}$$

$$[D] = \frac{E(1 - v)}{(1 + v)(1 - 2v)} \begin{bmatrix} 1 & \dfrac{v}{1 - v} & 0 \\ \dfrac{v}{1 - v} & 1 & 0 \\ 0 & 0 & \dfrac{1 - 2v}{2(1 - v)} \end{bmatrix} \tag{6.10}$$

Although the stress-strain relationship for the triangular elements has been shown, the stress-strain relationship for the rectangular finite elements is modified. The rectangle has four nodes and thus eight degrees of freedom are used in the displacement function resulting in an increase in the strain terms as shown in equation (6.11).

$$\varepsilon_x = \frac{\partial U_x}{\partial x} = \alpha_2 + \alpha_4 y \qquad \varepsilon_y = \frac{\partial U_y}{\partial y} = \alpha_7 + \alpha_8 x \tag{6.11}$$

$$\gamma_{xy} \quad = \quad \frac{\partial U_x}{\partial y} + \frac{\partial U_y}{\partial x} \quad = \quad \alpha_3 + x\alpha_4 + \alpha_6 + y\alpha_8$$

In PAFEC-FE the membrane elements are also isoparametric elements. That is, the two-dimensional membrane elements can have curved sides instead of straight sides. (In the case of three-dimensional elements, both the edges and faces can be curved). This complicates the element but essentially the constitutive equations are not affected.

6.6 Three-dimensional elements with three degrees of freedom per node

The stress–strain relationship for three-dimensional finite elements is an extension of the two-dimensional case. There are three degrees of freedom per node, U_x, U_y and U_z. Each node is at a corner of the three-dimensional brick element as shown in Figure 6.6, which gives eight nodes per element.

A displacement function with twenty-four degrees of freedom is required. The linear strain-displacement equation is as given in equation (6.12) and the constitutive matrix is shown in equation (6.13).

$$\varepsilon_x = \frac{\partial u}{\partial x}, \quad \varepsilon_y = \frac{\partial v}{\partial y}, \quad \varepsilon_z = \frac{\partial w}{\partial z},$$

$$\gamma_{xy} = \frac{\partial u}{\partial y} + \frac{\partial v}{\partial x}, \quad \gamma_{yz} = \frac{\partial u}{\partial z} + \frac{\partial w}{\partial y}, \qquad (6.12)$$

$$\gamma_{zx} = \frac{\partial w}{\partial x} + \frac{\partial u}{\partial z}$$

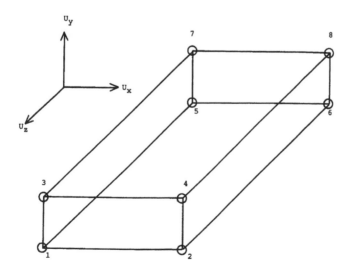

Figure 6.6 Eight node brick element with three degrees of freedom per node.

$$[D] = \begin{bmatrix} a & & & \text{symmetric} \\ b & a & & & & \\ b & b & a & & & \\ 0 & 0 & 0 & c & & \\ 0 & 0 & 0 & 0 & c & \\ 0 & 0 & 0 & 0 & 0 & c \end{bmatrix} \qquad (6.13)$$

where

$$a = \frac{E(1-v)}{(1+v)(1-2v)}, \quad b = \frac{av}{1-v}, \quad c = \frac{E}{2(1+v)}$$

Although an 8-noded brick-shaped element has been described, brick elements with 20 and 32 nodes are available. Wedge-shaped finite elements with 6, 15 and 24 nodes are also available.

6.7 Plate bending elements

In PAFEC-FE the plate bending elements are facet shell elements (e.g. type 44200) and are in an isoparametric formulation (allows curved sides), with displacement assumptions. In terms of the element axes, the matrices relating to the displacements U_x and U_y are exactly the same as those used for the isoparametric plane stress element type 36210. The bending matrices are based on the usual thin plate theory using the isoparametric transformation. For rectangular elements with four nodes, there are the two in-plane degrees of freedom U_x and U_y. The three additional degrees of freedom due to plate bending are as shown in Figure 6.7.

Equation (6.14) gives the simplest displacement function associated with plate bending finite elements.

$$w = \alpha_1 + x\alpha_2 + y\alpha_3 + x^2\alpha_4 + xy\alpha_5 + y^2\alpha_6 + x^3\alpha_7 + x^2y\alpha_8$$
$$+ xy^2\alpha_9 + y^3\alpha_{10} + x^3y\alpha_{11} + xy^3\alpha_{12} \qquad (6.14)$$

The strain vector at any point in the element is as shown in equation (6.15).

$$-\frac{\partial^2 w}{\partial x^2} = -[2\alpha_4 + 6x\alpha_7 + 2y\alpha_8 + 6xy\alpha_{11}]$$

$$-\frac{\partial^2 w}{\partial y^2} = -[2\alpha_6 + 2x\alpha_9 + 6y\alpha_{10} + 6xy\alpha_{12}] \qquad (6.15)$$

$$\frac{2\partial^2 w}{\partial x \partial y} = 2[\alpha_5 + 2x\alpha_8 + 2y\alpha_9 + 3x^2\alpha_{11} + 3y^2\alpha_{12}]$$

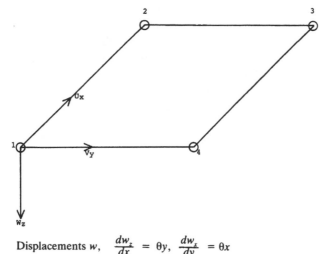

Displacements w, $\dfrac{dw_z}{dx} = \theta y$, $\dfrac{dw_z}{dy} = \theta x$

Figure 6.7 Plate bending degrees of freedom.

Equation (6.16) is the constitutive matrix.

$$\frac{Et^3}{12(1-v^2)}\begin{bmatrix} 1 & -v & 0 \\ -v & 1 & 0 \\ 0 & 0 & \tfrac{1}{2}(1-v) \end{bmatrix} \tag{6.16}$$

The information given in equations (6.14)–(6.16) is from the normal theory of bending of plates and includes all the assumptions inherent in that theory. For example, it is assumed that the plate is thin compared with its length and that its deflection is small compared with its thickness.

In the normal plate bending theory, the output stresses are the two bending moments plus a twisting moment per unit length of perimeter. However, in the PAFEC-FE system the transverse shear deflection may also be included, thus allowing thicker plates to be considered.

6.8 Shell elements

Many general engineering problems use plate-like members, with a thickness of 10% or less of the dimensions in the other two directions. For very thin plates, membrane effects only are important. For thicker plates both membrane and bending effects need to be considered. A series of plate elements may meet in such a way that bending in one plate is often coupled to both the bending and the membrane effects in adjacent elements. To overcome this effect it is necessary to incorporate in a single flat element, both membrane and bending effects. This can be accommodated in one of two ways, either by joining together as separate elements a membrane element and a plate bending element or by developing a thin shell element.

In the joining together as separate flat elements the membrane element and the plate bending element, the five degrees of freedom shown in Figure 6.8 are used. The elements incorporate the constitutive equations of sections 6.3 and 6.6.

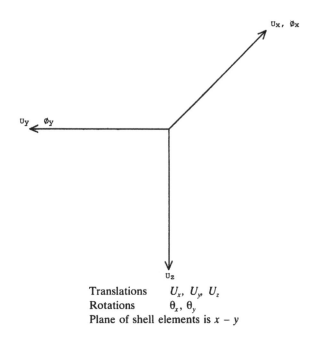

Translations U_x, U_y, U_z
Rotations θ_x, θ_y
Plane of shell elements is $x - y$

Figure 6.8 The five degrees of freedom per node for thin shell finite elements.

For the isoparametric, thick facet shell elements (e.g. type 45210), each of the five degrees of freedom per node, that is, three translations U_x, U_y and U_z plus the two rotations ϕ_x and ϕ_y are allowed to vary independently over the element surface. This is different to the thin shell facet element where the in-plane displacements are assumed to be two polynomial functions, with an additional single polynomial assumption for bending.

As the element can be used for thick homogeneous or sandwich plates, there are three stress–strain relationships, which are expressed through an 8×8 matrix shown in equation (6.17).

$$[\sigma] = [D] [\varepsilon] \qquad\qquad (6.17)$$

Table 6.1

	The three constitutive relationships		
Matrix terms in $[D]$	1	2	3 Column 2 terms plus those below
(1,1), (2,2)	$Eh/(1 - v^2)$	$2 E_f h_f (1 - v_f^2)$	$Eh/(1 - v^2)$
(1,2), (2,1)	$vEh/(1 - v^2)$	$2 v_f E_f h_f (1 - v_f^2)$	$vEh/(1 - v^2)$
(3,3)	$Eh/2(1 + v)$	$E_f h_f (1 - v_f)$	$Eh/2(1 + v)$
(4,4), (5,5)	$Eh^3/12(1 - v^2)$	$E_f (h + h_f)^2 h_f/2(1 - v_f^2)$	$E_f h_f^3/6(1 - v_f^2) + Eh^3/12(1 - v^2)$
(4,5), (5,4)	$vEh^3/12(1 - v^2)$	$v_f E_f (h + h_f)^2 h_f/2(1 - v_f^2)$	$vE_f h^3_f/6(1 - v_f^2) + Eh^3/12(1 - v^2)$
(6,6)	$Eh^3/24(1 + v)$	$E_f (h + h_f)^2 h_f/8(1 + v_f)$	$E_f h^3_f/12(1 + v_f) + Eh^3/24(1 + v)$
(7,7), (8,8)	$5Eh/12(1 + v)$	$Eh/2(1 + v)$	0

Core material properties E, v, h
Facing material properties E_f, v_f, h_f.

8×8 $[D]$ matrix terms are shown in Table 6.1.
The three constitutive relationships are as follows:

(1) The plate is homogeneous and isotropic, with the properties as defined in equation (6.17).
(2) The plate is of conventional sandwich construction with two equal thin faces and a flexible core. The facings and the core are both homogeneous.
(3) The plate is a sandwich construction as described above. Additional strain energy terms are added due to bending of each of the facings and the core and also due to the stretching of the core.

Two generally curved 'semi-loof' shell elements, one rectangular, the other triangular, are available in PAFEC-FE (types 43216 and 43215). Shear deflection is not included and thus both elements can only be used for thin shells. There are the usual three translatory degrees of freedom U_x, U_y and U_z. The rotations ϕ_x and ϕ_y are the rotations at the mid-side nodes and are called the 'loof' rotations. The constitutive matrix is shown in equation (6.18).

$$[D] = \frac{Eh}{1 - v^2} \begin{bmatrix} 1 & & & & & \\ v & 1 & & & \text{symmetric} & \\ 0 & 0 & \frac{1}{2}(1-v) & & & \\ 0 & 0 & 0 & h^2/12 & & \\ 0 & 0 & 0 & vh^2/12 & h^2/12 & \\ 0 & 0 & 0 & 0 & 0 & h^2(1-v)/24 \end{bmatrix} \qquad (6.18)$$

6.9 Large displacements

When strains, displacements and rotations are small, simple linear relationships between strains and first derivatives of displacement are used. If any of the quantities become large, then the strains are re-defined, for example, as shown in equation (6.19).

$$\varepsilon_{xx} = \frac{\partial U_x}{\partial x} + \frac{1}{2}\left[\left(\frac{\partial U_x}{\partial x}\right)^2 + \left(\frac{\partial U_y}{\partial x}\right)^2 + \left(\frac{\partial U_z}{\partial x}\right)^2 \right]$$

$$\varepsilon_{xy} = \frac{\partial U_x}{\partial x} + \frac{\partial U_y}{\partial x} + \left[\left(\frac{\partial U_x}{\partial x} \times \frac{\partial U_x}{\partial y}\right) + \left(\frac{\partial U_y}{\partial x} \times \frac{\partial U_y}{\partial y}\right) + \left(\frac{\partial U_z}{\partial x} \times \frac{\partial U_z}{\partial y}\right) \right]$$

$$(6.19)$$

Similar expressions are used for the other four strains but are not shown here.

6.10 Buckling

Although PAFEC-FE is only interested in the small displacement buckling, the non-linear strain expressions are required. The strain energy expression is based upon the strains being small and linearly related to the stresses, although the value of the individual terms which make up a strain may themselves be large. Care also has to be taken in the evaluation of the work done by the external loads as the direction of load application may alter as the displacements become large. The solution will be incremental with the load steps being small enough to require no iteration. At any stage of the calculation, a record is kept of the total displacement and stress.

6.11 Non-linear materials

It is usual to assume that the stiffness matrix is constant during the application of the load. This is the usual assumption of linearity. For many problems the non-linear effects cannot be neglected and have to be solved by assuming that either the stiffness matrix or the load vector are dependent upon the displacement vector.

6.11.1 Geometric non-linearity

In the case of geometric non-linearity or large displacements, as the load is applied, the stiffness matrix may change. Where the effects of stresses on the stiffness matrix are small, the load is divided into a series of sufficiently small increments which are applied one at a time. After the application of each load increment, the deflection is calculated using the usual linear theory of equations. The original coordinates of the nodes are then changed by the values of the deflections that have been calculated and the stiffness matrix re-calculated for the deformed structure. This process is repeated. Often, a large displacement solution can be found by incrementing the load.

6.11.2 Elastic-plastic behaviour

Elastic-plastic behaviour can be simply dealt with. The material is assumed to behave elastically before yield. After yielding, additional plastic strains occur and PAFEC-FE uses the linear solution modified with an incremental and iterative approach. Often, the stress-strain curve will approximate to the two straight lines of linear elasticity and linear strain hardening after yielding, so that it is defined by the elastic constants, the yield stress and the plastic stress-strain gradient. The plastic parts of the strain component contribute nothing towards the stresses and form the residual strains after unloading. Before yielding the plastic parts of the strain are zero.

In PAFEC-FE, the elastic-plastic constitutive relations are divided into two parts. The elastic parts of the strain components are related to the stresses by Hooke's law as shown in equation (6.20).

$$\varepsilon_{xx} + \frac{1}{E}\left[\sigma_{xx} - \nu(\sigma_{yy} + \sigma_{zz})\right]$$

$$\varepsilon_{xy} = \frac{1+\nu}{E}\sigma_{xy}\cdot 2$$

(6.20)

The plastic stress–strain relations use the Prandtl–Reuss equations associated with the von Mises yield criterion and give the relation of the plastic strain increment to the stresses as shown in equation (6.21).

$$\delta\varepsilon_{xxpl} = \delta\lambda\,(\sigma_{xx} - s)$$

$$\delta\varepsilon_{xypl} = \delta\lambda\,\sigma_{xy}\cdot 2$$

(6.21)

where

$$s = \frac{1}{3} [\sigma_{xx} + \sigma_{yy} + \sigma_{zz}] \qquad (6.21)$$

6.11.3 Creep behaviour

Creep of a material occurs when loads are applied over a long period of time. The resulting stresses, although they are well below the yield value, cause some permanent deformation. The mechanism is essentially that of a long term plastic flow and the theory is similar to elastic-plastic behaviour. The total strains are made up of elastic strains, ε_t, creep strains ε_{cr}, plastic strains ε_{pl} and thermal strains ε_{th}, which are related to the total stress through equation (6.22). However, PAFEC-FE does not include plastic and thermal strains in creep behaviour.

$$[\sigma] = [D][(\varepsilon_t) - (\varepsilon_{cr}) - (\varepsilon_{pl}) - (\varepsilon_{th})] \qquad (6.22)$$

PAFEC-FE assumes that all creep deformation will obey small strain theory and the properties used in generating the original stiffness matrix are not appreciably altered. Starting with the initial elastic solution using the applied loads and temperature (PAFEC-FE does not include temperature in creep behaviour), the subsequent incremental solutions are found by moving forward over each step and changing the stress fields. It is necessary to know the initial stress and total creep strain accumulated at the beginning of each time increment and these are stored. At the end of each time increment, the stress levels and accumulated creep strains are updated to the new values and the calculations are repeated over the next time increment.

6.11.4 Friction, gaps and non-linear springs

Sometimes parts of a structure are not rigidly jointed. They may compress each other, cause friction, cause tension or move apart. This is a non-linear problem and separate nodes are chosen at either side of a possible gap. The solution is an iterative one, which requires the derivation of a reduced stiffness matrix.

6.12 Thick shells of revolution

Thick shells of revolution, with axisymmetric loading are dealt with in a similar way to the two-dimensional isoparametric elements described in section 6.3. As the loading is axisymmetric the displacements U_x and U_y are all that are needed. For the completely axisymmetric case, these two displacements give rise to only four strains as shown in equation (6.23).

$$[\varepsilon] = [\varepsilon_{xx}\ \varepsilon_{yy}\ \varepsilon_{xy}\ \varepsilon_{zz}] \tag{6.23}$$

The stress vector is related to the strain vector as shown in equation (6.24).

$$[\sigma] = \begin{bmatrix} \sigma_{xx} \\ \sigma_{yy} \\ \sigma_{xy} \\ \sigma_{zz} \end{bmatrix} = \frac{E}{(1+v)(1-2v)} \begin{bmatrix} 1-v & v & 0 & v \\ v & 1-v & 0 & v \\ 0 & 0 & \frac{1}{2}-v & 0 \\ v & v & 0 & 1-v \end{bmatrix} [\varepsilon] \tag{6.24}$$

When the structure is axisymmetric, but the loading is not, then PAFEC-FE replaces the load by a number of Fourier components. However, the stress vector is related to the strain vector in exactly the same way as previously defined for the three-dimensional situation as given in equation (6.13)

6.13 Hybrid elements

The hybrid finite element method is more complicated and less frequently used than the displacement method. An assumption is made for $[\sigma]$ the vector of stresses or stress resultants and a further assumption is made for the displacement patterns on the element boundaries. Both the strains and the constitutive equations relating strains to stresses are obtained in the usual manner.

6.14 Temperature distribution

All temperature elements used in PAFEC-FE are isoparametric and therefore the geometrical descriptions and transformations are the same as those given in sections 6.4 and 6.5 for two- and three-dimensional structural elements.

6.15 Fracture mechanics

For the two-dimensional crack tip elements, a standard isoparametric element is used throughout the mesh. The element is distorting in the region of the crack tip, such that when an element with one mid-side node is used at a crack tip, the mid-side nodes should be moved from their usual position at the centre of each side to the quarter position as shown in Figure 6.9.

Although the elements at the crack tip have reasonably accurate stiffnesses, their local stress and displacement values generally have poor accuracy due to the inaccurate crack tip load modelling. The effect of this

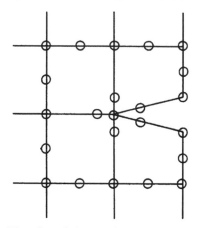

Figure 6.9 Distortion of elements in the region of a singularity.

inaccurate load modelling dies out after the greater distance of (a) or (b):

(a) the distance of one element or,
(b) the usual 'Saint-Venant' distance.

For further details of fracture mechanics, see chapter 11.

6.16 Anisotropic materials

Anisotropic materials possess directional properties. In order for PAFEC-FE
to present a generalized Hooke's law for anisotropic materials it is assumed
that the material is homogeneous and that all the components of strain are
linearly related to all the components of the stress. The strains are related
through a fourth rank material compliance tensor and a mathematical stress
tensor which can be reduced to a contracted compliance matrix as shown
in equation (6.25).

$$\varepsilon_i = C_{ij}\, \sigma_j \ (\ i, j = 1, \ldots, 6) \tag{6.25}$$

The C_{ij} matrix contains 36 terms, but due to symmetry and Maxwell's
reciprocal theorem, there are only nine independent terms as shown in equation (6.26).

$$[C] = \begin{bmatrix} C_{11} & C_{12} & C_{13} & 0 & 0 & 0 \\ & C_{22} & C_{23} & 0 & 0 & 0 \\ & & C_{33} & 0 & 0 & 0 \\ & & & C_{44} & 0 & 0 \\ & & & & C_{55} & 0 \\ \text{symmetric} & & & & & C_{66} \end{bmatrix} \tag{6.26}$$

The relationships between the terms in 'C' and the elastic constants of the material are as given in equation (6.27).

$$C_{11} = \frac{1}{E_{11}}, \qquad C_{22} = \frac{1}{E_{22}}, \qquad C_{33} = \frac{1}{E_{33}}$$

$$C_{12} = \frac{-v_{12}}{E_{11}}, \qquad C_{23} = \frac{-v_{23}}{E_{22}}, \qquad C_{31} = \frac{-v_{31}}{E_{33}} \qquad (6.27)$$

$$C_{44} = \frac{1}{G_{23}}, \qquad C_{55} = \frac{1}{G_{13}}, \qquad C_{66} = \frac{1}{G_{12}}$$

where E = Young's modulus, G = shear modulus, v = Poisson's ratio. 11 = x direction, 22 = y direction, 33 = z direction; 1,2 = xy plane, 1,3 = xz plane and 2,3 = yz plane.

For a transversely isotropic material which is an orthotropic material isotropic about one axis, the nine terms of the compliance matrix reduce to five independent terms as shown in equation (6.28).

$$[C] = \begin{bmatrix} C_{11} & C_{12} & C_{12} & 0 & 0 & 0 \\ & C_{22} & C_{23} & 0 & 0 & 0 \\ & & C_{22} & 0 & 0 & 0 \\ & & & (C_{22} - C_{23}) & 0 & 0 \\ & \text{symmetric} & & & C_{66} & 0 \\ & & & & & C_{66} \end{bmatrix} \qquad (6.28)$$

For a square symmetric material, where there are equal amounts of reinforcement in one plane, then the compliance matrix has only six independent terms as shown in equation (6.29).

$$[C] = \begin{bmatrix} C_{11} & C_{12} & C_{13} & 0 & 0 & 0 \\ & C_{11} & C_{13} & 0 & 0 & 0 \\ & & C_{33} & 0 & 0 & 0 \\ & & & C_{44} & 0 & 0 \\ & & & & C_{44} & 0 \\ & & & & & C_{66} \end{bmatrix} \qquad (6.29)$$

Many more forms of anisotropic material stress–strain equations are available.

6.17 Conclusion

The basis of the PAFEC-FE stress–strain relationships has been presented.

In essence, PAFEC-FE assumes a linear stress–strain relationship, with a number of variations. The detail of each possible relationship, will be found in PAFEC-FE 'Theory'. This chapter has concentrated on a series of the simpler elements followed by a description of the non-linear work, with the amount of detail being increasingly reduced as the chapter has progressed.

Reference

O.C. Zienkiewicz, *The Finite Element Method*, 4th edn., Vol. 1, McGraw-Hill, London (1989).

7 Two- and three-dimensional beam elements: framework analysis

7.1 Introduction

In the previous chapter we considered the mathematics behind some of the PAFEC-FE elements and in particular the various types of beam elements. We now look again at those beam elements and identify the circumstances in which they might be used. Whereas Chapter 1 used a simple beam structure to illustrate the finite element method we now consider more general beam structures. In particular we consider the PAFEC-FE analysis of a three-dimensional crane structure.

7.2 Details of the PAFEC-FE beam elements

7.2.1 Tension bar element 34400

The straight, uniform, tension bar element, 34400, carries end loads only and its formulation is exact in statics. It is used for structures with either pin-jointed members or beams with negligible bending and torsional stiffness. The element has two nodes each with one degree of freedom related to the displacement parallel to the beam length. After axis transformation, there are three degrees of freedom, U_x, U_y and U_z at each node. The axial force is assumed to be constant along the length. The data output is the axial force in the beam.

For this element, information under some of the following headers of the BEAMS module might be included.

SECTION.NUMBER   MATERIAL.NUMBER   AREA

7.2.2 Simple beam element 34000

Element 34000 is a straight uniform beam element with two nodes. The beam is for bending in two principal directions, axial forces and twisting about its shear centre. There are six degrees of freedom, U_x, U_y, U_z, ϕ_x, ϕ_y and ϕ_z at each of the two nodes, giving 12 degrees of freedom per element. The element may have any cross-section described by the second moments of area I_{yy} and I_{zz}, the area A and the torsional constant C. For a tubular beam, any two of the following three quantities are required; the inner diameter, the outer diameter or the wall thickness. The beam principal axes orientation is required and is added to the BEAMS module through AXIS or BETA or NODE.NUMBER.

For this element, information under some of the following headers of the BEAMS module might be included:

SECTION.NUMBER MATERIAL.NUMBER IYY IZZ AXIS.NUMBER
BETA TORSIONAL CONSTANT AREA NODE.NUMBER
INSIDE.DIAMETER OUTSIDE.DIAMETER THICKNESS

The element is used for framed structures and beam stiffeners where the flexural centre of the stiffener coincides with the line between the two end nodes. The axial force and twisting moment are constant along the length. The element formulation is exact in statics. The beam should not be used when the total length of a beam is less than five times the largest cross-section dimension. The cross-section is not permitted to warp. Shear and flexural centres must coincide.

Phase 9 of the PAFEC-FE program produces the shear forces, bending moments, axial force and torsional moment applied to the beam at its nodes and specified intermediate positions.

7.2.3 Shear deformation and rotary inertia beam element 34100

Element 34100 is a straight, uniform beam element which includes shear deformation. It caters for bending in two directions, an axial force and a twisting moment. The cross-section is described as in the BEAMS module (section 7.2.2), but includes the terms KY and KZ, which are used to correct for non-uniformity of shear stress over a cross-section.

The element is used for frames and stiffeners to plate, shell and box structures. The element assumes that shear forces, twisting moments and axial forces are all constant along the element length. The cross-section is not permitted to warp and the shear centre and the flexural centres must coincide. The total beam length must not be more than twenty times the element cross-section. There are six degrees of freedom, U_x, U_y, U_z, ϕ_x, ϕ_y and ϕ_z at each of the two nodes, giving 12 degrees of freedom per element.

The stress output of Phase 9 gives the shear forces, bending moments, axial force and torsional moment applied to the beam at its nodes and specified intermediate positions. Additionally, if the section moduli headers ZY and ZZ are used in the BEAMS module, then stresses are also calculated.

7.2.4 Simple beam element with offset 34200 and 34500

Elements 34200 and 34500 are straight beam elements with a six degree of freedom node at each end, making 12 degrees of freedom per element. 34200 is connected to the structure by means of offset nodes which define the position of the beam axis. If these node numbers are omitted from TOPOLOGY in the ELEMENTS module, then the beam is not offset. 34500 can be used where the two nodes of the topology will be the positions at which the beam is connected to the structure. The precise positions of the 34500 beam are given by V1 and V2 in the OFFSETS module. The cross-section is described in the beams module. The offsets at the two ends of the beam may be different.

The elements are for use with framed structures and stiffened plates in which the beam axis does not coincide with the line connecting the end nodes. Complex sections are built up by combining offset elements. The element formulation is exact in statics and includes bending in two directions, axial force and twisting. The axial force and twisting moments are constant. Shear deformation is not included and the element cross-section is not permitted to warp. The element should not be used when the total length of a beam is less than five times the largest cross-section dimension.

The stress output of Phase 9 gives the shear forces, bending moments, axial force and torsional moments applied to the beam at its nodes and at specified intermediate positions.

7.2.5   Shear deformation and rotary inertia beam element with offset 34600 and 34700

Elements 34600 and 34700 are straight uniform beam elements with a six degree of freedom node at each end, making 12 degrees of freedom per element. 34600 is connected to the structure by means of offset nodes which define the position of the beam axis. If these node numbers are omitted from TOPOLOGY in the ELEMENTS module, then the beam is not offset. 34700 can be used where the two nodes of the TOPOLOGY will be the positions at which the beam is connected to the structure. The precise positions of the 34700 beam are given by V1 and V2 in the OFFSETS module. The cross-section is described in the beams module. The offsets at the two ends of the beam may be different.

The elements are for use with framed structures and stiffened plates in which the beam axis does not coincide with the line connecting the end nodes. Complex sections are built up by combining offset elements. The element formulation is exact in statics and includes bending in two direc-tions, axial force and torsional moment. The shear forces, axial force and torsional moments are constant along the beam length. The element cross-section is not permitted to warp. The total beam length must not be more than twenty times the element cross-section and the shear and flexural centres must coincide.

The data output gives the shear forces, bending moments, axial force and torsional moments applied to the beam at its nodes and at specified intermediate positions.

7.2.6   Shear deformation and rotary inertia beam element with offset and a reduced number of degrees of freedom 34210

Element 34210 is a straight uniform beam element with a three degree of freedom node U_x, U_z and ϕ_y at each end of the shear centre, making 6 degrees of freedom per element. It is connected to the structure by means of offset nodes which are termed as non-structural nodes in PAFEC-FE. If these node numbers are omitted from the TOPOLOGY entry in the ELEMENTS module, then the beam is not offset. The offsets at the two ends of the beam may be different.

The element is found to be useful in a limited number of cases, such as skin and stringer problems in the aerospace industry. The element allows bending in one direction and axial force. Shear deformation is taken into account in the stiffness matrix formulation and rotary inertia in the element mass matrix. The degrees of freedom allow stretching along the length of the beam (U_x) and bending about the principal y axis of the beam (U_z, ϕ_y).

The BEAMS module is used to input the details and the relevant headers are:

     SECTION.NUMBER   MATERIAL.NUMBER   IYY   AXIS.NUMBER
     BETA   AREA   NODE.NUMBER   KY

The element cross-section is not permitted to warp. The total beam length must not be less than five times the largest cross-section dimension and the shear and flexural centres must coincide. Bending is allowed in one plane only. PAFEC-FE output gives the axial force, z shear force and the *MY* bending moments applied to the actual beam ends in element axes.

7.2.7   Curved beam element 34300

Element 34300 includes shear deformation, rotary inertia and forms part of a circle. There are six degrees of freedom at each of the two structural nodes making 12 degrees of freedom per element. The element is used for curved members in frame structures and also for curved stiffeners attached to shell structures.

The element in-plane bending is coupled with stretching of the element and the rigid body and constant stress states are fully modelled. The out-of-plane and torsional motion are also treated in the same way and are uncoupled with the in-plane motions. The constituent mass matrix which includes the effect of rotary inertia is derived. The element must not be used for straight beams and must not subtend an angle of less than about 5° or greater than about 45°. The principal axes of the cross-section must be perpendicular and parallel to the plane of the element. The output from Phase 9 includes the forces and moments applied to each end of the beam in the radial, tangential and axial directions.

The element is specified by the two node numbers in the TOPOLOGY, which are the positions at the centres of the area of the cross-section at the two ends of the element. The NODE.NUMBER entry in the BEAMS module gives the centre of curvature of the element. IZZ relates to bending in the plane of the element. The AXIS.NUMBER and BETA entries in the BEAMS module are ignored.

7.3 Warnings and errors for beam elements

The warning and error checks that are made in PAFEC-FE depend upon the method used to define the beam principal axes.

For a straight uniform beam, an offset beam and a shear beam three checks are made as follows:

(a) AXIS method: In order to define the X,Y, X,Z principal bending planes, only the direction of the Y axis is used. The system takes the X axis of the beam (node 1 to node 2) and the Y axis of AXIS.NUMBER in order to define an X,Y plane. If the Y axis of AXIS.NUMBER is parallel to the beam, an X,Y plane is not defined. Thus the program checks that the Y direction of AXIS.NUMBER is not within 5° (warning message) or 1° (error message) of the beam axis.

(b) BETA method: The principal X,Y bending plane is defined relative to the global axes and as the beam is viewed from node 1 to node 2, the X,Y plane appears as an edge. BETA is the angle by which the edge needs to be rotated about the X axis in order for it to lie in a plane parallel to the global X,Y plane passing through node 1. This method cannot be used if the axis of the beam lies along a parallel to the global Z plane. The program checks that the beam axis is not within 5° (warning message) or 1° (error message) of the global Z axis.

(c) NODE method: This method uses a third node in addition to the two end nodes (nodes 1 and 2) of the beam to define the X,Y principal bending plane. If the three nodes are co-linear, the NODE method cannot be used. Thus the program checks that the perpendicular distance from the NODE to the line of the beam axis is not less than 5% (warning message) or 1% (error message) of the beam length.

As stated before, a warning message from PAFEC-FE allows execution of the program to continue whereas an error message signals premature termination. Results obtained after warning messages have appeared should be regarded with caution. Ideally there should be a re-modelling to avoid such warnings. In this and other cases warnings may be indicative of a decrease in numerical accuracy.

For a curved beam the definition of the radii needs careful consideration and a check is made to ensure that:

(a) The subtended angle at the centre of curvature is less than 45°. An error message is signalled if the angle subtended by the element is less than 3° or greater than 50°.

(b) The distance between the two end nodes and the centre of curvature, indicated by NODE.NUMBER, is checked to ensure the distances are within 1% of each other. An error message is signalled if they differ by more than 1%.

7.4 Two-dimensional and three-dimensional frame analysis

The individual elements of a two-dimensional, pin-jointed truss resist only tension and compression. Two-dimensional beams used in two-dimensional rigid jointed frames, withstand both bending moments and loads applied normal to their longitudinal axis. The result is axial forces, bending moments and shear. The pin-jointed truss nodal points are uniquely defined as the physical panel points, but for frames composed of beam elements, it is necessary to decide where precisely the nodal points are located. This calls for consideration of the anticipated behaviour of the total structure related to the properties of the beam element and becomes particularly important if the two-dimensional structure being analysed is a sub-structure of a larger three-dimensional structure.

In many structures, individual parts act independently of each other in that forces are transmitted between the parts, but no further interaction occurs. A structure may be decomposed into two-dimensional parts for analysis. Some analysts suggest this should almost always be done but it does require considerable experience to determine the correct two-dimensional sub-structures.

A three-dimensional analysis shows how the actual structure reacts, but two problems must be considered. Firstly, the linear equations to be solved are more extensive, which requires extra computing resources, and secondly, care must be taken to ensure the beam Y axis as defined in the AXIS or BETA entry in the BEAMS module does not give a principal axis direction parallel to the beam axis.

The structural integrity of the majority of three-dimensional structures depends upon three-dimensional interaction and thus should be analysed as three-dimensional structures.

7.5 The analysis of a three-dimensional crane structure

7.5.1 Model description

Figure 7.1 shows the general arrangement and node numbering of the three-dimensional crane structure with four splayed legs. Figure 7.2 shows a side and an end elevation, with Figure 7.3 showing the plan. In Figure 7.2, the origin of the axes are located directly below node 3 and on the $Y = 0$ plane, which includes nodes 1, 2, 6 and 10. The top of the crane is at $Y = 200$ cm. Centimetres are used rather than millimetres as centimetres are the units used in structural steelwork tables. Using Figures 7.1, 7.2 and 7.3, it is possible to build up the data input in modular form.

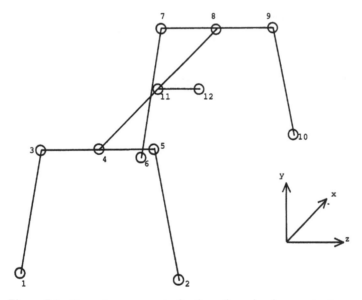

Figure 7.1 General arrangement of a three-dimensional crane structure.

7.5.2 PAFEC-FE data input

Looking now to the construction of a PAFEC-FE data set along the lines outlined in chapters 3 and 4, a suitable TITLE record would be:

TITLE   Three-dimensional crane structure

We now construct a nodes module. It is laid out so that the positions of the joints are readily indicated. The dimensions are assumed to be in centimetres. The material is a steel hollow box section and the section properties for steel are usually given by the manufacturers in centimetres.

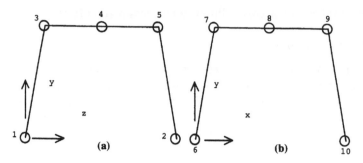

Figure 7.2   Elevation of the crane structure. (a) End elevation; (b) side elevation.

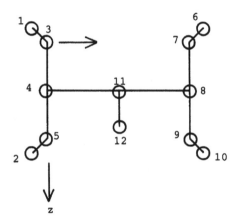

Figure 7.3   Plan of the crane structure.

NODES
AXIS.NUMBER = 1

NODE.NUMBER	X	Y	Z
1	−20	0	−20
2	−20	0	220
3	0	200	0
4	0	200	100
5	0	200	200
6	320	0	−20
7	300	200	0
8	300	200	100
9	300	200	200
10	320	0	220
11	150	200	100
12	150	200	175

We note that the THICKNESS header is not required since neither semi-loof nor thick shell elements are involved. The AXIS.NUMBER entry of 1 refers to the right-hand Cartesian axes set already programmed into PAFEC-FE. In our example, the origin is taken to be in the plane on which the structure stands and vertically below node number 3.

The next module is the ELEMENTS module. It is used to define the type of beam element being used, the element property and topology. The module is as follows:

```
ELEMENTS
ELEMENT.TYPE = 34000
 NUMBER   PROPERTIES   TOPOLOGY
    1         1          1    3
    2         1          2    5
    3         1          6    7
    4         1         10    9
    5         2          3    4
    6         2          4    5
    7         2          7    8
    8         2          8    9
    9         2         11   12
   10         3          4   11
   11         3         11    8
```

Element 34000 is suitable for frame structures and is satisfactory because although shear deformation is not included in its development, the minimum beam length of 750 cm is more than five times the largest cross-sectional dimension of 10 cm shown in Figure 7.4. The 11 elements of the structure are numbered 1 to 11 according to their position as shown in Figure 7.3. The PROPERTIES column has the values 1, 2 and 3 and relates to the beams with the TOPOLOGY shown. For example, the first four element numbers are 1, 2, 3 and 4. All have the property 1 and refer to the SECTION.NUMBER 1 in the BEAMS module. Each of the four elements refers to the legs of the crane as shown in Figures 7.1, 7.2 and 7.3. The elements with the PROPERTIES of value 2, are numbered 5, 6, 7, 8 and 9. They relate to the beams which run in the direction of the Z axis. The final two elements are numbered 10 and 11. They both have the PROPERTIES value of 3.

The next module is the BEAMS module in which physical characteristics of the beams are specified. Note that PAFEC-FE does not require the beams' external dimensions as shown in Figure 7.4

```
BEAMS
AXIS.NUMBER = 1
 SECT   MATE   IYY   IZZ   TORS   AREA   ZY    ZZ
   1     11    474   474   761    35.5   94.9  94.9
   2     11    283   283   439    18.9   56.6  56.6
   3     11    234   234   361    15.3   46.8  46.8
```

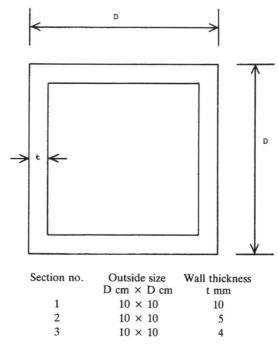

Section no.	Outside size D cm × D cm	Wall thickness t mm
1	10 × 10	10
2	10 × 10	5
3	10 × 10	4

Figure 7.4 Square hollow section member and size.

For the BEAMS module, three sizes of square hollow steel sections are to be used. Their dimensions and properties are readily found from a wide range of published design tables. Figure 7.4 shows a cross-section through SECTION.NUMBER = 1. The constant property line has been used after the module name to state that for all elements the AXIS.NUMBER = 1. Three methods are available in PAFEC-FE to define the principal axes of the section. The methods are, AXIS.NUMBER method, BETA method and the third node method. Only the AXIS.NUMBER method is used in this chapter.

As is often the case not all the available headers are required. The first is SECTION.NUMBER which we may abbreviate to SECT. An entry under SECT refers to the PROPERTIES column in the ELEMENTS module where three different properties are specified, namely 1, 2 and 3. The outside dimensions of the three section sizes are all the same and are 10 cm by 10 cm. The wall thicknesses of 1 cm, 0.5 cm and 0.4 cm are shown in Figure 7.4, but are not explicitly shown in the data. The thicknesses are implicitly incorporated within the values of IYY, IZZ and AREA. The next input is the MATERIAL.NUMBER, which is specified as 11 and refers to the MATERIAL module. In this case although mild steel is one of the standard materials included in PAFEC-FE, its unit of length (in PAFEC-FE) is metres. In our case we use centimetres as the data input unit of length, therefore the Young's modulus has to be changed from PAFEC-FE units of 209E9 N/m^2 to 209E5 N/cm^2. In steelwork design the units of force per unit area, that is pressure, is expressed in N/m^2, etc. and not MPa as in SI units.

The next two terms in the contents line are IYY and IZZ which are the two principal second moments of area of the section. Again the units must be consistent and as the section is square and symmetrical the two values are the same at 474 cm^4 for section number 1. The following term is the TORSIONAL.CONSTANT and has units of cm^4. Care must be taken to ensure that the PAFEC-FE defined torsional constant symbol of C is directly equivalent to the steel designers torsional constant J which is again in cm^4. Another torsional constant C is often used by steel designers, but has units of cm^3.

Take care to ensure that a defined value is used for the TOR-SIONAL.CONSTANT as PAFEC-FE sets a default value of IYY + IZZ, which is only true for circular sections. AREA defines the cross-sectional area of the section and for section number 1 is 35.5 cm^2. As the beams are not curved and the axis number is 1, then NODE.NUMBER is ignored and given a default value of 0. Element 34000 does not include shear and the shear constants KY and KZ for the cross-sections are not required.

For circular sections, PAFEC-FE requires two of the following three entries to be defined, INSIDE.DIAMETER, OUTSIDE.DIAMETER and THICKNESS. For solid sections only, the OUTSIDE.DIAMETER is required. Again care must be taken, as the PAFEC-FE manual refers to pipes and tubular beam elements and not to circular sections. Many engineering publications classify tubular sections as being hollow sections of any shape and if, for example as in this chapter, values are inserted for INSIDE.DIAMETER, OUTSIDE.DIAMETER and THICKNESS, then PAFEC-FE will ignore the input values and use its own calculated values. Thus for non-circular tubular sections, e.g. square hollow sections, the INSIDE.DIAMETER, OUTSIDE.DIAMETER and THICKNESS are ignored and assigned the default values of 0.

The final two items in the contents list are ZY and ZZ, the sectional elastic moduli, which for section number 1 have the values of 94.9 cm^3 and 94.9 cm^3. The values of ZY and ZZ are calculated in the usual way as in the bending of beams.

The next module is MATERIAL. Once more the Young's modulus value has to be changed from 209E9 N/m^2 to 209E5 N/cm^2. This change is introduced through this module and as a statics analysis is required only the values of Young's modulus E equal to 209E5 N/cm^2 and Poisson's ratio NU equal to 0.30 are specified as the MATERIAL.NUMBER = 11. The data input module MATERIAL is as follows:

MATERIAL

MATERIAL.NUMBER	E	NU
11	209E5	0.30

At present, the three-dimensional crane structure is not secured in space and if a load is applied, would move as a rigid body. To secure the crane such that it will resist loads, the RESTRAINTS module is used as follows:

RESTRAINTS
PLANE = 0
AXIS.NUMBER = 1

NODE.NUMBER	DIRECTION
1	12346
2	12346
6	12346
10	12346

Two constant property items have been chosen. Firstly, AXIS.NUMBER = 1, which states that a right-handed Cartesian coordinate set is being used. Secondly, PLANE = 0, which indicates that the restraint is applied to the nodes specified under NODE.NUMBER, which are nodes 1, 2, 6 and 10. The final term in the contents line is DIRECTION, which refers to the direction of the degree of freedom to be restrained. These are numbered 1, 2, 3, 4, 5 and 6. They refer to the directions U_x, U_y, U_z, and ϕ_x, ϕ_y, ϕ_z, respectively. For this example, the base of the beams are secured to the ground such that no movement is allowed in directions 1, 2, 3, 4 and 6. Only direction 5, that is ϕ_y, is not restrained.

The final data input module, before the END.OF.DATA marker, is LOADS and is as follows:

LOADS

CASE.OF.LOAD	NODE.NUMBER	DIRECTIONS. OF.LOADS	VALUE.OF.LOADS
1	12	2	−10000

END.OF.DATA

One load case is being considered, thus CASE.OF.LOAD is 1. The load is applied at NODE.NUMBER 12. As in the RESTRAINTS module, the numbers 1, 2, 3, 4, 5 and 6 refer to the directions U_x, U_y, U_z, ϕ_x, ϕ_y and ϕ_z, respectively. The load is applied in the Y direction and is shown as DIRECTION.OF.LOADS equal 2. The VALUE.OF.LOAD is 10000 N in the negative Y direction. The end of the data input is signified by END.OF.DATA.

The full data input file, including an IN.DRAW for data verification and an OUT.DRAW for the displaced shape, is as follows:

TITLE   THREE DIMENSIONAL CRANE STRUCTURE
CONTROL
C No special options
CONTROL.END

NODES
AXIS.NUMBER = 1

NODE.NUMBER	X	Y	Z
1	−20	0	−20
2	−20	0	220
3	0	200	0
4	0	200	100
5	0	200	200
6	320	0	−20
7	300	200	0
8	300	200	100
9	300	200	200
10	320	0	220
11	150	200	100
12	150	200	175
100	400	400	400

C Use Node 100 for viewing the structure
IN.DRAW

TYPE	INFO	NODE
2	3	100

OUT.DRAW

PLOT.TYPE	NODE
1	100

ELEMENTS
ELEMENT.TYPE = 34000

NUMBER	PROPERTIES	TOPOLOGY	
1	1	1	3
2	1	2	5
3	1	6	7
4	1	10	9
5	2	3	4
6	2	4	5
7	2	7	8
8	2	8	9
9	2	11	12
10	3	4	11
11	3	11	8

BEAMS
AXIS.NUMBER = 1

SECT	MATE	IYY	IZZ	TORS	AREA	ZY	ZZ
1	11	474	474	761	35.5	94.9	94.9
2	11	283	283	439	18.9	56.6	56.6
3	11	234	234	361	15.3	46.8	46.8

MATERIAL

MATERIAL.NUMBER	E	NU
11	209E5	0.30

RESTRAINTS
PLANE = 0

```
AXIS.NUMBER = 1
NODE.NUMBER   DIRECTION
      1            12346
      2            12346
      6            12346
     10            12346
LOADS
CASE.OF.LOAD   NODE.NUMBER   DIRECTIONS   VALUE
      1              12            2        -10000
END.OF.DATA
```

In Phase 1 of the PAFEC-FE program, the above data modules are read in and validated. Each module is checked to ensure that the necessary column headers have an entry and that the entry does not conflict with other supplied data. The node numbers are also checked to ensure they are positive and that each node and its number are unique. The data modules input is reproduced in order under the headings: BEAMS, ELEMENTS, GLOBAL COORDINATES, LOADS, MATERIALS and RESTRAINTS.

The output under the heading GLOBAL COORDINATE identifies each node, but as no extra nodal generation has been involved, the global coordinate data output is the same as the data input in the NODES module. What is more important is the output under the headings TABLE SHOWING ELEMENTS ATTACHED TO NODES and DEGREE OF FREEDOM NUMBERS AT NODES. This allows identification of the data. The elements attached to nodes output which appears at Phase 4 are as follows:

TABLE SHOWING ELEMENTS ATTACHED TO NODES

NODE	ELEMENTS	NODE	ELEMENTS	NODE	ELEMENTS
1	1	2	2	3	1 5
4	5 6 10	5	2 6	6	3
7	3 7	8	7 8 11	9	4 8
10	4	11	9 10 11	12	9

The above data output can be cross checked against the input data in the ELEMENTS module.

The nodal degree of freedom numbers as shown below, are also produced at Phase 4. A star indicates a constraint. The output can be used to check that the correct restraint requirements have been inserted.

DEGREE OF FREEDOM NUMBERS AT NODES

NODE NUMBER	1	2	3	4	5	6	7	8	9	10	11	12
UX D.O.F.	*	*	3	9	15	*	22	28	34	*	41	47
UY D.O.F.	*	*	4	10	16	*	23	29	35	*	42	48
UZ D.O.F.	*	*	5	11	17	*	24	30	36	*	43	49
PHIX D.O.F.	*	*	6	12	18	*	25	31	37	*	44	50
PHIY D.O.F.	1	2	7	13	19	21	26	32	38	40	45	51
PHIZ D.O.F.	*	*	8	14	20	*	27	33	39	*	46	52

For example, node 1 is shown as being restrained in all directions except PHIY.

Phase 7 gives the nodal displacements in centimetres and the nodal rotations in radians. As the crane is symmetrical about a constant U plane through nodes 11 and 12, only half the crane structure need be checked. It should be pointed out that because the model is symmetrical, the data output will be symmetrical, and this should be validated. For example the displacement results of nodes 5 and 9 should be the same, but with the positive and negative signs interchanged for UX, PHIY and PHIZ. The displacements are in global coordinates.

NODE NUMBER	UX	UY	UZ	PHIX	PHIY	PHIZ
5	0.0731	−0.0147	−0.0612	−0.842	0.963	−1.248
9	−0.0731	−0.0147	−0.0612	−0.842	−0.963	1.248

The translations are in centimetres and the rotations have been multiplied by 1E3. A further check is made on nodes 11 and 12 which are expected to have the largest movements as node 12 is the loaded node. The data output is as follows:

NODE NUMBER	UX	UY	UZ	PHIX	PHIY	PHIZ
11	−0.0	−0.6761	−0.0774	20.53	0.0	−0.0
12	−0.0	−2.4533	−0.0774	25.28	0.0	−0.0

After determining the nodal displacements and rotations, the beam element stresses, forces and moments are calculated in the local beam axes as shown in Figure 7.5. The sign convention is that negative bending moment about an element axis denotes sagging with respect to that axis. The key to the output is as follows:

S = SHEAR FORCE
M = BENDING MOMENT
X = DISTANCE ALONG THE BEAM
S = DM/DX

AXIAL = DIRECT STRESS
AVETAUY = AVERAGE SHEAR STRESS IN THE Y DIRECTION
AVETAUZ = AVERAGE SHEAR STRESS IN THE Z DIRECTION
BENSIGY = BENDING STRESS IN THE Y DIRECTION
BENSIGZ = BENDING STRESS IN THE Z DIRECTION
CFSTMAX = MAXIMUM COMBINED FIBRE STRESS
CFSTMIN = MINIMUM COMBINED FIBRE STRESS

The members with the largest slenderness ratio are elements 1, 2, 3 and 4, with 55.3 and a length of 201.99 cm. The element having the least slenderness ratio of 19.4 is element 9 which is 75 cm long. It must be remembered that the lengths of the members in PAFEC-FE are from nodal point to nodal

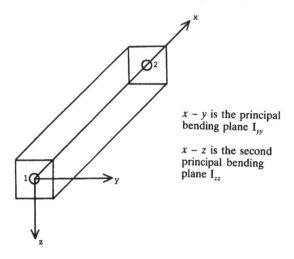

$x - y$ is the principal bending plane I_{yy}

$x - z$ is the second principal bending plane I_{zz}

Figure 7.5 Planes and axes for the beam element.

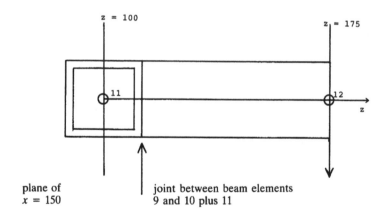

Figure 7.6 Position of beam element 9 related to nodes 11 and 12.

point and that element 9 which connects node 11 to node 12 will in practical terms start at the edge of elements 10 and 11 which is 5 cm from node 11 as shown in Figure 7.6. As the displacements for nodes 5, 9 and 11 were given earlier, the stresses for elements 2 and 9 only will be printed and are shown below. The results for element 4 are the same as for element 2, but with some of the positive and negative signs interchanged.

ELEMENT NO	NODE NO	AXIAL FORCE	Y-SHEAR FORCE	Z-SHEAR FORCE	TORSIONAL MOMENT	Y-BENDING MOMENT
2	2	−4.547E3	4.851E2	6.014E2	−3.141E3	1.934E4
	5	−4.547E3	4.851E2	6.014E2	−3.141E3	1.021E5

| 9 | 11 | 0.314 | −1.000E4 | 0.0 | 0.0 | 0.0 |
| | 12 | 0.314 | −1.000E4 | 0.0 | 0.0 | 0.0 |

ELEMENT NO	NODE NO	Z-BENDING MOMENT	SLENDER RATIO	AXIAL	AVETAUY	AVETAUZ
2	2	1.203E4	55.3	−1.281E2	1.367E1	1.694E1
	5	1.100E5	55.3	−1.281E2	1.367E1	1.694E1
9	11	7.500E5	19.4	0.0166	−5.291E2	0.0
	12	1.069E1	19.4	0.0166	−5.291E2	0.0

ELEMENT NO	NODE NO	BENSIGY	BENSIGZ	CFSTMAX	CFSTMIN
2	2	−2.038E2	1.268E2	−4.587E2	2.025E2
	5	1.076E3	1.159E3	2.364E3	2.107E3
9	11	0.0	1.325E4	1.325E4	−1.325E4
	12	0.0	0.1888	0.2054	−0.1722

Looking more closely at the beam element 9, the downward applied load of 10000 N at node 12 at the end of element 9 shows as a Y-SHEAR FORCE of −1E4 at nodes 11 and 12, which is correct. The Z-SHEAR FORCE, TORSIONAL MOMENT and the Y–BENDING MOMENT are correctly given as zero. The Z-BENDING MOMENT, AVETAUY, BENSIGZ, CFSTMAX and CFSTMIN are also correctly given. The relationship between AXIAL FORCE and AXIAL STRESS is also correct.

The above results for beam 9 using element 34000, indicate that the calculated values are correct. This suggests that the results for the other elements are also likely to be correct. It is worth noting that the results for beam 9 conform to good engineering intuition and experience. If doubts existed regarding the results, then a different element could be used (see section 7.6) or the mesh could be refined (see section 7.7). It must be remembered that 34000 is a simple beam element and does not include shear deformation. It would be useful to solve the three-dimensional crane example using element 34100 which does include shear deformation.

7.6 Element 34100 and the three-dimensional crane structure

The data input is exactly the same as for the previous example, but with one alteration. The line ELEMENT.TYPE = 34000 in the ELEMENTS module is changed to ELEMENT.TYPE = 34100.

The inclusion of shear deflection in this element means that the numerical accuracy may be impaired, if the element length is more than about twenty times the element cross-sectional dimension. This would be the case for all beams longer than 200 cm. That is, beam element numbers 1, 2, 3 and 4. Thus the displacements and stresses at node 5 for the two elements will be compared and the stresses for element 9 of the two beam elements 34000 and 34100 will also be compared. The deflections for the two elements at nodes 5 and 12 are shown below:

NODE NUMBER	UX	UY	UZ	PHIX	PHIY	PHIZ
5 [34000]	0.0731	−0.0147	−0.0612	−0.842	0.963	−1.248
5 [34100]	0.0735	−0.0146	−0.0604	−0.836	0.960	−1.254
12 [34000]	−0.0	−2.4533	−0.0774	25.28	0.0	−0.0
12 [34100]	−0.0	−2.4674	−0.0766	25.29	0.0	−0.0

The UY displacement for element 34100 is larger than for 34000, which is as would be expected for a beam in pure bending, as the inclusion of shear deformation increases its displacement. However, the variations are very small.

We turn now to the stresses at node 5 for element 2 and nodes 11 and 12 for element 9. The stresses for element 34100 are compared with element 34000. If the values for element 34100 are the same as element 34000, then a blank is left in the table as follows:

ELEMENT NO	NODE NO	AXIAL FORCE	Y-SHEAR FORCE	Z-SHEAR FORCE	TORSIONAL MOMENT	Y-BENDING MOMENT
2 [34000]	5	−4.547E3	4.851E2	6.014E2	−3.141E3	1.021E5
[34100]	5	−4.542E3	4.822E2	5.981E2	−3.190E3	1.015E5
9 [34000]	11	0.314	−1.000E4	0.0	0.0	0.0
[34100]	11					
9 [34000]	12	0.314	−1.000E4	0.0	0.0	0.0
[34100]	12					

ELEMENT NO	NODE NO	Z-BENDING MOMENT	SLENDER RATIO	AXIAL	AVETAUY	AVETAUZ
2 [34000]	5	1.100E5	55.3	−1.281E2	1.367E1	1.694E1
[34100]	5		55.3	−1.279E2	1.358E1	1.685E1
9 [34000]	11	7.500E5	19.4	0.0166	−5.291E2	0.0
[34100]	11					
9 [34000]	12	1.069E1	19.4	0.0166	−5.291E2	0.0
[34100]	12	0.956E1				

ELEMENT NO	NODE NO	BENSIGY	BENSIGZ	CFSTMAX	CFSTMIN
2 [34000]	5	1.076E3	1.159E3	2.107E3	−2.364E3
[34100]	5	1.070E3		2.101E3	−2.356E3
9 [34000]	11	0.0	1.325E4	1.325E4	−1.325E4
[34100]	11				
9 [34000]	12	0.0	0.1888	0.2054	−0.1722
[34100]	12		0.1689	0.1856	−0.1523

The above results show that for beam element 9, which is loaded in pure bending, the stress results are almost identical for elements 34000 and 34100. When the complexity of the loading increases, then the stress results from the two elements begin to differ.

7.7 Increased degrees of freedom

In the finite element method, there is a relationship between the number of degrees of freedom, the complexity of the finite element used and the accuracy of the answer, whether displacements or stresses. Assuming the correct finite element has been used to model the required structure, in this case beam elements, then increasing the complexity of the element used from 34000 to 34100 should increase the accuracy of the displacement and stress results. Also increasing the number of elements used again increases displacement accuracy and stress result accuracy.

In the following example the three-dimensional crane structure is again analysed, but with additional beam elements and nodes in the legs as shown in Figure 7.7 and in the following replacement NODES and ELEMENTS modules.

NODES
AXIS.NUMBER = 1

NODE. NUMBER	X	Y	Z
1	−20	0	−20
2	−20	0	220
3	0	200	0
4	0	200	100
5	0	200	200
6	320	0	−20
7	300	200	0
8	300	200	100
9	300	200	200
10	320	0	220
11	150	200	100
12	150	200	175
13	−10	100	−10
14	−10	100	210
15	310	100	−10
16	310	100	210

ELEMENTS
ELEMENT.TYPE = 34000

NUMBER	PROPERTIES	TOPOLOGY	
1	1	1	13
2	1	2	14
3	1	6	15

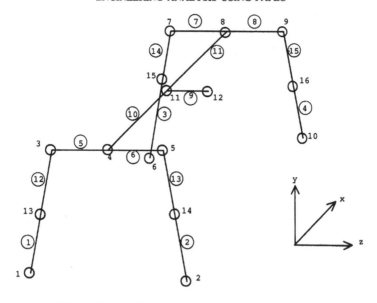

Figure 7.7   Additional nodes and elements in the legs.

4	1	10	16
5	2	3	4
6	2	4	5
7	2	7	8
8	2	8	9
9	2	11	12
10	3	4	11
11	3	11	8
12	1	13	3
13	1	14	5
14	1	15	7
15	1	16	9

Although the preceding data set is quite correct, the reader will be aware that the nodes and elements have been constructed on an individual basis. Whilst this is acceptable for the present problem it would be impractical for larger problems. In such cases the labour saving methods of chapter 4 should be used. As an illustration we show two alternative approaches, both producing the same model.

(1) The LINE.NODES module may be used to generate the extra nodes on the legs of the model. This removes the necessity to specify coordinates. The NODES module would only specify nodes 1 to 12 but the data would include the following module:

```
LINE.NODES
LIST
   1   13    3
   2   14    5
   6   15    7
   9   16   10
```

The ELEMENTS module follows, as before.

(2) The previous approach still requires elements to be specified individually. This may be overcome by using the PAFBLOCKS module to generate the beam elements along each of the legs of the crane. In this case the NODES module would once again only specify nodes 1 to 12 and the ELEMENTS module would specify elements 5 to 11. The remaining elements, on the legs of the model would be specified using the following modules:

```
PAFBLOCKS
TYPE = 6
ELEMENT.TYPE = 34000
PROPERTIES = 1
N1 = 1
BLOCK.NUMBER   TOPOLOGY
   1               1   3
   2               2   5
   3               6   7
   4               9  10
MESH
REFERENCE   SPACING.LIST
   1             2
```


This latter approach has the distinct advantage of requiring just one number, the number under the SPACING.LIST header of the MESH module to be changed in order to change the number of elements per leg of the model.

The displacement results for nodes 5 and 12 using element 34000 with the four columns as two elements are the same as when one element, 34000, is used for each leg. With respect to the stresses, there is again no difference in results between one element and two elements in the legs.

Performing the analysis with two beam elements in each leg instead of one beam element, but using element 34100 in both cases, will give the same displacement results, but a few of the stress results will change. However, changes where they occur are not significant. A blank space in the following table indicates an unchanged value.

ELEMENT NO	NODE NO	AXIAL FORCE	Y-SHEAR FORCE	Z-SHEAR FORCE	TORSIONAL MOMENT	Y-BENDING MOMENT
13 [1 ELM]	5	−4.547E3	4.851E2	6.014E2	−3.141E3	1.021E5
[2 ELM]	5		4.852E2			
9 [1 ELM]	11	0.314	−1.000E4	0.0	0.0	0.0
[2 ELM]	11					
9 [1 ELM]	12	0.314	−1.000E4	0.0	0.0	0.0
[2 ELM]	12					

ELEMENT NO	NODE NO	Z-BENDING MOMENT	SLENDER RATIO	AXIAL	AVETAUY	AVETAUZ
13 [1 ELM]	5	1.100E5	55.3	−1.281E2	1.367E1	1.694E1
[2 ELM]	5		27.6			
9 [1ELM]	11	7.500E5	19.4	0.0166	−5.291E2	0.0
[2 ELM]	11					
9 [1 ELM]	12	1.069E1	19.4	0.0166	−5.291E2	0.0
[2 ELM]	12	1.425E1				

ELEMENT NO	NODE NO	BENSIGY	BENSIGZ	CFSTMAX	CFSTMIN
13 [1 ELM]	5	1.076E3	1.159E3	2.107E3	−2.364E3
[2 ELM]	5			2.108E3	
9 [34000]	11	0.0	1.325E4	1.325E4	−1.325E4
[34100]	11				
9 [1 ELM]	12	0.0	0.1888	0.2054	−0.1722
[2 ELM]	12		0.2518	0.2684	−0.2352

As both element 34000 and 34100 have different formulations, it is to be expected that there will be some differences between their results. However, as both elements are exact in statics, it is to be expected that increasing the number of the same type of beam elements would make minimal difference to the displacements and stresses.

7.8 Crane with vertical legs

In the previous example, the crane legs were not vertical, but splayed out at an approximate angle of 8° to the vertical. The reason for choosing this angle is that for beam elements, PAFEC-FE checks that the beam Y axis as defined in the AXIS or BETA entry in the BEAMS module does not give a principal axis direction parallel to the beam axis. If the angles are less than 5° a warning is signalled and if less than 1° an error is signalled. To overcome this restriction and to use vertical legs the AXIS.NUMBER, BETA or the NODE.NUMBER methods in the BEAMS module as described in section 7.3 are used.

7.9 Conclusions

The straight beam elements 34000, 34100, 34200, 34500, 34600, 34700 and 34210 plus the curved beam element 34300 have been described. The straight beam elements are for use in frame structures, beam stiffeners to plate and in shell and box structures, both with and without the beam axis coinciding with the line connecting the end nodes. The curved beam element is used for curved stiffeners attached to shell structures.

The beam elements 34000 and 34100 have been used to solve a three-dimensional crane structure. The data input is given and the output discussed. A major advantage of using finite element beam elements rather than the more traditional stiffness method of analysis, is that the finite element beams can join with a vast array of other finite elements in PAFEC-FE to analyse complex structures.

8 Three-dimensional elements

8.1 Introduction

PAFEC-FE uses a series of straight edged, as well as isoparametric (that is elements that allow curved edges and curved sides) three-dimensional brick, triangular prism and tetrahedral elements for three-dimensional stress analysis. Wherever possible a model using two-dimensional rather than three-dimensional elements is to be preferred. The saving in data preparation and computer resources can be considerable.

8.2 The need for three-dimensional elements

The need for a three-dimensional finite element stress analysis rather than a two-dimensional finite element stress analysis appears to relate to a trade-off between data preparation effort and computer time against the accuracy and acceptability of the results. Even with mesh generation, data preparation and error checking in three dimensions can be time consuming. Generally, three-dimensional stress analysis is more accurate than two-dimensional analysis, but a combination of two- and three-dimensional analysis, with sub-structuring and mesh refinement can be very fruitful. The use of higher order three-dimensional elements, with more nodes and degrees of freedom per element, reduces the number of required elements, reduces the data input and increases the accuracy of the stresses and displacements when related to the number of elements used.

Two-dimensional finite element stress analysis makes one of two assumptions regarding stress or the strain in the third direction. It assumes that either the stress is zero, the 'plane stress' assumption, or the strain is zero, the 'plane strain' assumption. For the remainder of this chapter, it is accepted that the variation of stresses in all directions is non-linear with the result that three-dimensional finite elements must be used.

8.3 Three-dimensional elements available in PAFEC-FE

8.3.1 General description

A series of isoparametric, three-dimensional finite elements are available.

Each element has three translatory degrees of freedom U_x, U_y and U_z at each node. The elements are not limited in their application except that if they are to be used with non-three-dimensional elements, then advice from more experienced PAFEC-FE users is required. As explained in section 4.5.1 a warning is given if there is element distortion, with an error message in extreme cases.

Elements are available in three basic shapes: a brick element, a triangular prism element and a tetrahedral element. The brick elements are more accurate than the triangular prism elements, but the triangular prism elements are easier to fit into complex geometries. The tetrahedral elements are less accurate than even the triangular prism elements, but can be used to mesh the more difficult finite element models that require other than brick or triangular elements. The tetrahedral elements should either be avoided or only used far away from areas of interest. All three shapes of element are isotropic, but orthotropic elements are available and are obtained from the isotropic elements by replacing the final digit of the element type by a 5, for example, 37120 and 37125.

For isotropic elements, the principal stresses are given at each of the nodes and also in some cases at the centres of some of the faces and at the centre of the element. For the orthotropic elements, stresses in principal material directions are given. It is usual to check the stress discontinuities between elements meeting at a node to check on the overall accuracy of the finite element model.

8.3.2 Twenty node, twelve edge, brick element

The generally shaped three-dimensional twenty node brick element shown in Figure 8.1 has six curvilinear faces with twelve edges. There are eight corner nodes and one mid-side node on each of the edges making 24 degrees of freedom per element. PAFEC-FE numbers the nodes and defines the directions N1, N2 and N5 as shown in Figure 8.1.

The element may distort a reasonable amount from the basic cubical shape. If the mid-side node numbers are omitted from the TOPOLOGY entry in the ELEMENTS module, then the corresponding sides are taken as straight. For the isotropic element 37110 the PROPERTIES entry in the ELEMENTS or PAFBLOCKS modules gives the material number. For the orthotropic elements, the PROPERTIES entry refers to the LAMINATES module, where the ORTHOTROPIC.MATERIAL and material principal directions are specified.

Pressure loading may be applied to all faces of the element provided that all corner nodes on the loaded face have non-zero pressure values.

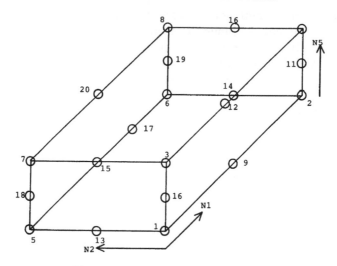

Figure 8.1 Twenty node brick element.

8.3.3 Further twelve edge brick elements

A series of brick elements with twelve edges and six faces are available as follows:

(a) Eight noded element types 37100 and 37105 have 24 degrees of freedom. A typical element is shown in Figure 8.2.

(b) Eight noded element types 37120 and 37125 have two nodes along each side, making a 32 node, 96 degree of freedom element.

(c) Element types 37130 and 37135 have 16 nodes and 48 degrees of freedom. Two opposite faces have a mid-side node on all four edges. There are no mid-side nodes in the third direction, which make the element useful when the stresses vary less in one direction than in the other two directions.

(d) Element types 37140 and 37145 have 24 nodes and 72 degrees of

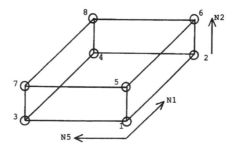

Figure 8.2 Eight node brick element.

freedom per element. Two opposite faces have two mid-side nodes on all four edges. There are no mid-side nodes in the third direction. The element may be used where the stress variation is less in one direction than in the other two directions. The aspect ratio of the lengths of the longest sides with two third-point nodes to the shortest sides with no mid-side nodes can be up to twelve, which is larger than for elements type 37110 and 37115.

(e) Element types 37150 and 37155 have 28 nodes and 84 degrees of freedom. There are eight corner nodes and two third point nodes, except along four quasi-parallel sides, where there is one mid-side node.

(f) Element types 37160 and 37165 have two faces each with four nodes. The two faces are connected by curved edges, each with one mid-side node, making 12 nodes and 36 degrees of freedom per element.

8.3.4 Triangular prism elements

(a) The simplest triangular prism elements are type 37200 and 37205 which have straight edges. They have 6 nodes and 18 degrees of freedom, but the displacement assumption is simple and the stress results leave much to be desired.

(b) Element types 37210 and 37215 have 6 corner nodes and 9 mid-edge nodes making a total of 15 nodes and 45 degrees of freedom. These elements are the most accurate triangular prism elements for general three-dimensional stress analysis, but are not as accurate as the brick elements type 37110 and 37115. Figure 8.3 shows element type 37210.

(c) Element types 37220 and 37225 are similar to 37210 and 37215 but

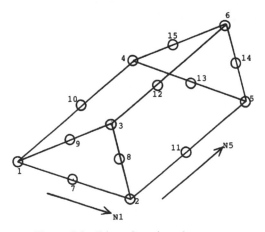

Figure 8.3 Triangular prism element.

have two mid-side nodes on each of their edges making 24 nodes and 72 degrees of freedom.

(d) Element types 37230 and 37235 have two triangular faces with 6 nodes on each face. The faces are joined by quadrilateral faces with no mid-side nodes on the opposite edges. The element has 12 nodes and 36 degrees of freedom. The element is used where the stresses vary less rapidly in one direction compared with the other two directions, for example when it is used in conjunction with brick element type 37120 for thick shell problems.

(e) Element types 37240 and 37245 are similar to 37230 and 37235, but have two third point nodes on the edges of the two triangular faces, making 18 nodes and 54 degrees of freedom. They are for thick shell analysis in conjunction with brick element types 37140 and 37145.

(f) Element types 37250 and 37255 have two triangular faces each with two third point nodes on the three edges. There is one mid-side node on each of the three lines connecting the triangular faces giving 21 nodes and 63 degrees of freedom. These elements are suitable for use with the brick element types 37150 and 37155.

(g) Element types 37260 and 37265 have two triangular faces with 5 nodes on each face. The faces are joined by straight edges with no mid-side nodes. The elements have 10 nodes and 30 degrees of freedom. They are suitable for use with brick element types 37160 and 37165.

8.3.5   Tetrahedral elements

(a) Element types 37300 and 37305, as shown in Figure 8.4, have four triangular faces, four nodes and 12 degrees of freedom.

(b) Element types 37310 and 37315 have a node at each vertex and at the mid-point of each edge. The elements have 10 nodes and 30 degrees of freedom.

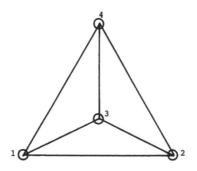

Figure 8.4   Tetrahedral element.

8.4 Data validation for three-dimensional elements

It is essential that data input mistakes are noticed quickly and then corrected. PAFEC-FE in Phase 1, goes through a validation of the input data. If errors are found, they are printed out at the end of the phase, together with the reasons and the analysis is stopped. If PAFBLOCKS are used, then checks are made again at the end of Phase 2. The first check is on the validity of the node numbers chosen by the user. The second check concerns the element shape. Mild distortions will cause a warning and severe distortions will cause an error and stop the analysis.

For the three-dimensional elements, the warnings and errors related to element shape are given in Table 8.1. Each face is checked using the criteria in Table 8.1, with the exception that the check on H_{max} is omitted for the quadrilateral.

Table 8.1 Warning and error limits for three-dimensional[a] elements

Element shape	Warning		Error	
Triangle	$5 <$	$R_{max} < 15$	R_{max}	≥ 15
	$15 <$	$\theta_{min} < 30$	θ_{min}	≤ 15
	$150 <$	$\theta_{max} < 165$	θ_{max}	≥ 165
	$0.00001 <$	$H_{max} < 0.01$	H_{max}	≥ 0.01
Quadrilateral	$5 <$	$R_{max} < 15$	R_{max}	≥ 15
	$25 <$	$\theta_{min} < 45$	θ_{min}	≤ 25
	$135 <$	$\theta_{max} < 155$	θ_{max}	≥ 155

[a]R_{max} = length of the longest side divided by the length of the shortest side; θ_{min} = the minimum angle between chords across any adjacent element sides; θ_{max} = the maximum angle between chords across any adjacent element sides; H_{max} = the distance of the fourth node from the plane of the first three, divided by the maximum side length.

8.5 An application of three-dimensional finite elements

Three-dimensional brick elements are to be used to determine the maximum tensile stress and the maximum vertical displacement in a concrete slab placed on a box of soil as shown in Figure 8.5. The maximum downward stress on the top of the bottom layer is also to be found.

The applied load is a vertical single point load, which can be placed at any of the mesh cross points on the concrete slab as shown in Figure 8.6. The concrete slab is two metres square and symmetrically placed on the box of soil. Due to the lines of symmetry A–A, B–B, C–C and D–D, only 6 of the 25 load positions need to be considered.

The load will result in stresses which will vary, non-linearly in all three directions. Figure 8.7 shows a cross-section through the slab and the soil.

In this chapter we will choose the following system of units:

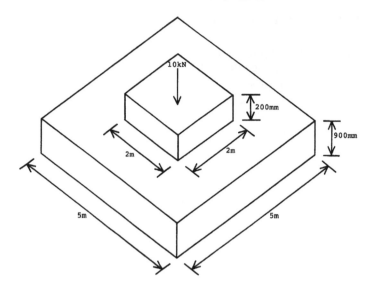

Figure 8.5 Concrete slab on a box of soil.

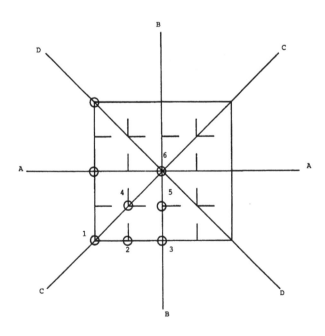

Figure 8.6 Load points on top of concrete slab.

(1) Length   metres (m)
(2) Force   newtons (N)
(3) Stress, modulus, pressure   N/m² (Pa)
(4) Translatory stiffness   N/m
(5) Moment   N m
(6) Density   kg/m³
(7) Mass   kg

We will introduce four new materials identified by numbers 11 to 14 to the PAFEC-FE program. The materials and their properties are as shown in Table 8.2.

Table 8.2 Materials and their properties used in the program

No.	Material	Thickness (mm)	Young's modulus	Poisson's ratio	Density (kg/m³)
11	Polystyrene layer	600	3E6	0.3	100
12	Soil sub-base layer	300	2E8	0.25	1800
13	Sand bedding layer	50	75E6	0.25	1500
14	Concrete raft	150	34E9	0.15	2400

8.5.1 Model description

From Figures 8.5, 8.6 and 8.7 it is possible to construct the data for the model. The steps involved have already been outlined in earlier chapters but we will nonetheless comment on each item in turn.

For a TITLE record we chose the following and append a few extra comment lines. Replacing the C of the comment records by an * would have caused the comments to be appended to the title.

```
TITLE   Concrete Slab Placed On Soil — Test Number 1
C
C   Slab size 2 m by 2 m by 150 mm
C
```

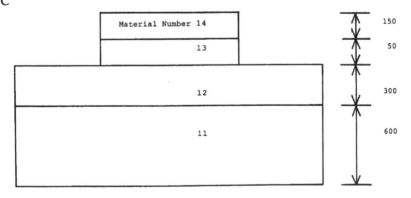

Figure 8.7 Section through slab and soil.

The NODES module is constructed along familiar lines. In this example we have chosen a right-handed Cartesian axes set with the X and Z axes in the horizontal and the Y axis in the vertical through the thickness of the model. The axes are shown in Figure 8.8. The origin of coordinates is set at node number 15, which is on the base, diagonally opposite node 2. The X axis runs in the direction of node 15 to node 16.

In Figure 8.8 the vertical axis Y has five values, namely, 0, 0.6, 0.9, 0.95 and 1.1. Both the X and the Y axis have the same four values of 0, 1.5, 3.5 and 5. It does not matter in what order the nodal positions are numbered, provided each node has a unique node number.

It will be seen that the two lower layers of material have common nodes 3, 4, 17 and 18 and are thus joined together. The two upper layers also have common nodes 9, 10, 13 and 14 and are thus joined together. There appear to be no common nodes between the two middle layers. However, through appropriate meshing using the PAFBLOCKS module and information supplied under the MESH header we will arrange for the two middle layers to have common nodes. The model may then be analysed as one structure. The NODES module, data input is as follows:

NODES
AXIS.NUMBER = 1

NODE.NUMBER	X	Y	Z
1	0	0	5
2	5	0	5
3	0	0.6	5
4	5	0.6	5
5	0	0.9	5
6	5	0.9	5
7	1.5	0.9	3.5
8	3.5	0.9	3.5
9	1.5	0.95	3.5
10	3.5	0.95	3.5
11	1.5	0.9	1.5
12	3.5	0.9	1.5
13	1.5	0.95	1.5
14	3.5	0.95	1.5
15	0	0	0
16	5	0	0
17	0	0.6	0
18	5	0.6	0
19	0	0.9	0
20	5	0.9	0
21	1.5	1.1	3.5
22	3.5	1.1	3.5
23	1.5	1.1	1.5
24	3.5	1.1	1.5

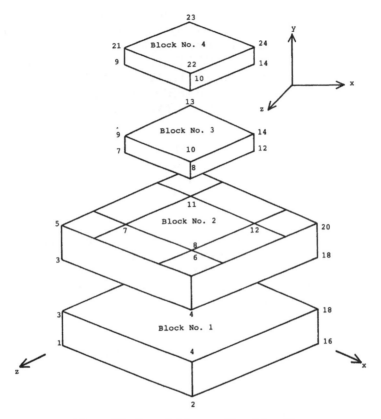

Figure 8.8 Separated layers with nodal positions.

We proceed to generate the eight noded brick elements within each of the four layers. At this stage we are establishing the type of the elements to be used and through information under the appropriate headers pointing to other modules which will define the material properties and the actual mesh spacing.

The PAFBLOCKS module is as follows:

```
PAFBLOCKS
ELEMENT.TYPE = 37100
BLOCK.NUMBER   PROPERTIES   N1  N2  N5   TOPOLOGY
    1             11          1   1   2    1   2   3   4  15  16  17  18
    2             12          1   1   2    3   4   5   6  17  18  19  20
    3             13          3   3   2    7   8   9  10  11  12  13  14
    4             14          3   3   2    9  10  21  22  13  14  23  24
```

We now establish material properties through the MATERIAL module. For our model the coefficient of thermal expansion ALPHA, the hysteretic damping factor MU, the thermal conductivity K, the specific heat SH and the bulk modulus BULK.MODULUS are not required and so the relevant headers are not included. The required input data are the Young's modulus E, the Poisson's ratio NU and the mass density RO. The ten standard material types already programmed into PAFEC-FE, numbered 1 to 10 are not to be used. We will define our material types 11, 12, 13 and 14 which will appear in the PROPERTIES column of the PAFBLOCKS module and so far refer to the MATERIAL module. The MATERIAL module is as follows:

MATERIAL

MATERIAL.NUMBER	E	NU	RO
11	3E6	0.3	100
12	2E8	0.25	1800
13	75E6	0.25	1500
14	34E9	0.15	2400

For all four materials, the values of mass density are readily available from simple laboratory experiments, as are the Young's modulus and Poisson's ratio, for materials 11 and 14. However, a series of assumptions must be made with respect to the Young's modulus and the Poisson's ratio values for materials 12 and 13.

For materials 12 and 13, which are the sub-base and the bedding sand, respectively, the value of the Young's modulus is usually determined from the California Bearing Ratio (CBR) value. As a number of relationships between the CBR and the Young's modulus are available, the Young's modulus used in this chapter, is stated and the reader can determine the CBR value from whichever relationship is chosen (Croney, 1977).

The Poisson's ratio for a sub-base and a sand can be used to model their degree of saturation. For totally dry materials 12 and 13, the Poisson's ratio will be 0.0. For completely saturated materials 12 and 13, the Poisson's ratio will be 0.5. However, due to numerical difficulties a value of 0.5 should not be used in PAFEC-FE. The maximum value to be used should be about 0.45. The MATERIAL module assumes that the materials 12 and 13 are only moderately wet and a value of 0.25 is used.

We now specify the mesh. The problem is to ensure that the sizes of the elements of the blocks are such that there is no overlapping. That is, with the exception of nodes on the surface, every other node is at a point at which four elements meet. This uniformity ensures a continuity in the mesh which in turn generally leads to more accurate analysis.

By way of an example we consider the X-Y plane of $Z = 3.5$. The mesh we wish to achieve is shown in Figure 8.9. Since we have a symmetrical structure a similar diagram would apply to the other three cross-sections. From the figure it can be seen that the ratio of the lengths of the larger elements to the smaller elements in the X-direction is 75 to 20, or 15 to 4 and that there are 14 elements in the X-direction in the lowest block and 10 in the top block. This X-direction corresponds to the N1 direction of the

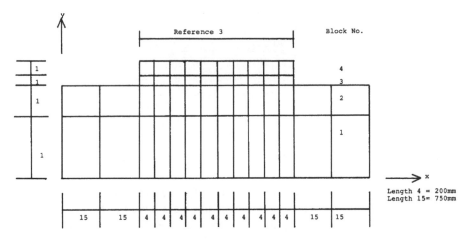

Figure 8.9 Section through mesh module at $Z = 3.5$.

PAFBLOCK and so we have a SPACING.LIST entry of 15 15 4 4 4 4 4 4 4 4 4 4 15 15 for the lower block, signifying a total of 14 elements to be spaced in the given ratio. Since the elements in the upper block are all identical we have a single SPACING.LIST entry of 10.

By the aforementioned symmetry we have identical entries referred to by the N2 header. The N5 header refers to a SPACING.LIST entry of 1, indicating a thickness in each block of just one element.

MESH
REFERENCE	SPACING.LIST
1	15 15 4 4 4 4 4 4 4 4 4 4 15 15
2	1
3	10

At this point we may mention that the particular mesh we have chosen is such that elements are all well proportioned in that they all conform to the recommended PAFEC-FE standards explained in section 4.5.1.

In order to verify the mesh we might, as suggested in section 4.5.3, wish to produce a drawing. This might be achieved by the following IN.DRAW module. The output is shown in Figure 8.10. We choose to view the model from the point [6.0, 2.0, 6.0]. To include this point in the IN.DRAW module it would have to be assigned a node number. We will assume that the NODES module has been extended to include a definition of node number 25 at this point. The fact that the node is a non-structural node does not matter. A suitable IN.DRAW module would be:

IN.DRAW
TYPE	INFO	ORIENTATION	NODE
3	0	1	25

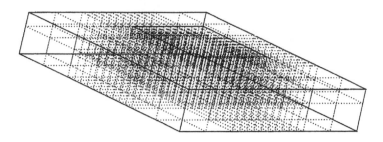

Figure 8.10 Output from the IN.DRAW module.

In this example we have requested a solid line boundary with a broken line interior. Node numbers and element numbers have not been requested. There are so many they might be difficult to see if only simple graphics hardware is available. The structure is to be viewed from node 25 and the X-axis is to be drawn in the conventional direction.

Having completed the details of how the model is to be meshed we now turn our attention to the specifications of the constraints and loadings.

At the present point of the data input, the model is free to move as a rigid body in space and some form of restraint is required. This is achieved through the use of the RESTRAINTS module as follows:

RESTRAINTS

NODE.NUMBER	PLANE	AXIS.NUMBER	DIRECTION
1	1	1	1
1	2	1	2
1	3	1	3
16	1	1	1
16	3	1	3

We now explain the construction of the module. From Figure 8.11, it can be seen that the concrete slab is on a soil foundation. The soil foundation is part of a larger soil mass and would normally be assumed to extend a considerable distance beyond the 5 × 5 m box size shown in Figure 8.11. However, the model represents a laboratory box of 900 mm high steel sides and with a flat concrete base, consequently, the model is to be restrained on the five faces bounded by the nodes as follows:

Face 1   Nodes   1, 2, 5, 6
Face 2   Nodes   2, 16, 6, 20
Face 3   Nodes   15, 16, 19, 20
Face 4   Nodes   1, 15, 5, 19
Face 5   Nodes   1, 2, 15, 16 (the base)

The first four faces are to be restrained horizontally and face 5, the base, is to be restrained vertically.

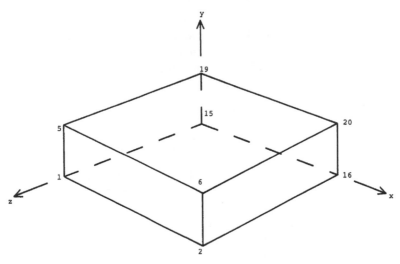

Figure 8.11 The four side faces and base that are restrained.

Consider the first line of data input.

The NODE.NUMBER is 1. This is the node [0, 0, 5] to which the restraint is to be applied. The PLANE is 1, which places a restraint to the plane of constant X ($X = 0$) through the NODE.NUMBER 1 in Cartesian coordinates, as the AXIS.NUMBER has been defined as 1. The plane of $X = 0$ through the nodal point 0, 0, 5 contains all the nodes in the plane defined by the corner nodal points 1, 5, 15, 19. The DIRECTION is defined as 1. This refers to the direction of the degree of freedom to be restrained. In this case 1 means in the X direction, allowing the soil box to be free to move in the other two directions Y and Z.

The output from Phase 4 of the PAFEC-FE program shows how the RESTRAINTS module has had effect. For example the following reflects the constraints implicit in the first line of the data module:

```
THE FOLLOWING 45 POINTS HAVE BEEN FOUND ON THE PLANE
FOR THE CASE   NODE   PLANE   AXIS   DIRECTION
                 1       1       1        1
NODES       1    3    5   15   17   19   40   55   70   85  100
115       130  145  160  175  190  205  220  261  276  291  306  321
336       351  366  381  396  411  426  441  482  497  510  525  540
555       570  585  600  615  630  645  658
```

As expected from preceding paragraphs not only are the co-planar nodes which we specifically defined in the NODES module included but also the co-planar nodes generated by the PAFBLOCKS and MESH modules. The co-ordinates of these generated nodes may be seen from the output of Phase 2 of this program.

The Phase 4 output also provides a means for checking that restraints have been applied correctly. For our problem the output would include the following table of degree of freedom numbers. An * indicates that a constraint has been applied and so no movement is allowed in the direction shown.

NODE NUMBER	1	2	3	4	5	6	7	8	9	10
UX D.O.F	*	*	*	*	*	*	5	8	11	14
UY D.O.F	*	*	1	2	3	4	6	9	12	15
UZ D.O.F	*	*	*	*	*	*	7	10	13	16

In the above it can be seen that node 1 is completely restrained whilst node 3 may only move in the Y-direction. On the other hand node 7 has complete freedom of movement.

We now wish to impose a load at the centre point of the topmost face. This point which has coordinates [2.5, 1.1, 2.5] will have to be given a node number so that it may be referred to by the LOADS module. For the purpose of our example we will assume that the earlier NODES module has been extended to ascribe a node number 26 to this point.

In passing we note two other ways of assigning node numbers which may not have been explicitly defined in the NODES module. We could either:

(a) Use the CONTROL option STOP to terminate execution at the end of Phase 2 and from the output ascertain which node number has been assigned to the point in question. The original data set would then be modified to include a load at this node.

(b) Use the REFERENCE.IN.PAFBLOCKS module as described in section 4.7 to prescribe a suitable node number of our choice at the required point. In the example shown, the node number 26 is given to the centre point of the topmost block by means of the following module.

REFERENCE.IN.PAFBLOCKS

BLOCK.NUMBER	C1	C2	C3	POSITION	NODE
4	5	5	1	8	26

Assuming that the centre point has been given a node number of 26 we may specify a downward load of 10 000 N at this point by the following module.

LOADS

CASE.OF.LOAD	NODE.NUMBER	DIRECTIONS.OF.LOAD	VALUE.OF.LOAD
1	26	2	−10000

The entry under the CASE.OF.LOAD header shows that only one load case is being considered. The load is specified as a negative quantity since the number 2 under the DIRECTION.OF.LOAD header refers to the positive Y direction.

For completeness we list the data we have constructed in the form of a complete PAFEC-FE data set, and where appropriate we have taken advantage of defaults and allowable abbreviations. The problem under consideration is a large one and so, as explained in section 3.5.2, it will be necessary specifically to allocate working storage.

```
TITLE Concrete Slab Placed on Soil — Job Number 1
C
C   Slab size 2 m by 2 m by 150 mm
C
CONTROL
PHASE = 4
BASE = 40000
PHASE = 7
BASE = 80000
PHASE = 9
BASE = 80000
CONTROL.END
NODES
AXIS.NUMBER = 1
NODE.NUMBER   X    Y     Z
     1        0    0     5
     2        5    0     5
     3        0    0.6   5
     4        5    0.6   5
     5        0    0.9   5
     6        5    0.9   5
     7        1.5  0.9   3.5
     8        3.5  0.9   3.5
     9        1.5  0.95  3.5
    10        3.5  0.95  3.5
    11        1.5  0.9   1.5
    12        3.5  0.9   1.5
    13        1.5  0.95  1.5
    14        3.5  0.95  1.5
    15        0    0     0
    16        5    0     0
    17        0    0.6   0
    18        5    0.6   0
    19        0    0.9   0
    20        5    0.9   0
    21        1.5  1.1   3.5
    22        3.5  1.1   3.5
    23        1.5  1.1   1.5
    24        3.5  1.1   1.5
C   Node for viewing the model
    25        6.0  2.0   6.0
C   Node to be loaded
    26        2.5  1.1   2.5
```

PAFBLOCKS
ELEMENT.TYPE = 37100

BLOCK.N	PROP	N1	N2	N5	TOPOLOGY							
1	11	1	1	2	1	2	3	4	15	16	17	18
2	12	1	1	2	3	4	5	6	17	18	19	20
3	13	3	3	2	7	8	9	10	11	12	13	14
4	14	3	3	2	9	10	21	22	13	14	23	24

MESH

REFERENCE	SPACING.LIST												
1	15	15	4	4	4	4	4	4	4	4	4	15	15
2	1												
3	10												

IN.DRAW

TYPE	INFO	ORIENTATION	NODE
3	0	1	25

MATERIAL

MATERIAL.NUMBER	E	NU	RO
C Polystyrene			
11	3E6	0.3	100
C Sub-base			
12	2E8	0.25	1800
C Bedding sand			
13	75E6	0.25	1500
C Concrete			
14	34E9	0.15	2400

RESTRAINTS

NODE.NUMBER	PLANE	DIRECTION
1	1	1
1	2	2
1	3	3
16	1	1
16	3	3

LOADS

NODE.NUMBER	DIRECTIONS.OF.LOAD	VALUE.OF.LOAD
26	2	−10000

END.OF.DATA

8.5.2 Discussion of the output

Phase 7 gives the displacement output and as expected the maximum U_y displacement is a compression at node 26. The value is −169.24E−6 m. The displacements for the centre node 26 and the four corner nodes 21, 22, 23 and 24 at the top of the concrete slab are as follows:

NODE NUMBER	TRANSLATIONS MULTIPLIED BY 1E6		
	UX	UY	UZ
26	0.002	−169.24	−0.000

21	2.243	-120.98	-2.245
22	-2.238	-120.98	-2.244
23	2.243	-121.01	2.242
24	-2.241	-121.00	2.243

As the model is symmetrical, the X and Z displacements for node 26 should be identical. The Y values for nodes 21, 22, 23 and 24 should also be identical. If the + and − signs for nodes 21, 22, 23 and 24 are ignored, then the X and Z values for nodes 21, 22, 23 and 24 should be the same. In all of the displacements that should be identical, the variation is less than 0.32%. This indicates that there may be a very small amount of numerical round off error, but it must be linked to the fact that the numerical model uses 2352 degrees of freedom. Thus 2352 linear simultaneous equations have been solved.

Following on from Phase 7 is Phase 9 the STRESS.ELEMENTS MODULE. The data output states the following:

(1) Element type: in this case element 37100
(2) The load case: No 1
(3) Global stresses: Sigma-X, Sigma-Y and Sigma-Z, the stresses in the global axes
(4) Principal stresses: Sigma-1 is the most positive principal stress and Sigma-2, the most negative principal stress
(5) AX, AY and AZ are the angles of Sigma-1 to the global axes BX, BY and BZ are the angles of Sigma-2 to the global axes and Sigma-3 is perpendicular to Sigma-1 and Sigma-2

Looking at node 744 [2.5, 0.95, 2.5], the node at the centre of the bottom of the concrete raft and elements 537, 538, 547 and 548 which are in the concrete raft, the stress output at node 554, because the model is symmetrical, should be the same for all the elements at that node, with the exception of the angles, which will alter depending upon the location of the element. The output is shown below.

ELEM NO.	NODE NO.	*****GLOBAL STRESSES*****			***PRINCIPAL STRESSES***		
		SIGMA-X	SIGMA-Y	SIGMA-Z	SIGMA-1	SIGMA-2	SIGMA-3
537	744	3.22E5	-1.97E5	3.22E5	4.84E5	3.22E5	-3.59E5
538	744	3.22E5	-1.97E5	3.22E5	4.84E5	3.22E5	-3.59E5
547	744	3.22E5	-1.97E5	3.22E5	4.84E5	3.22E5	-3.59E5
548	744	3.22E5	-1.97E5	3.22E5	4.84E5	3.22E5	-3.59E5

ELEMENT NO	VON.MISES STRESS	ANGS.OF.PRINCIPAL DIRECTIONS					
		AX	AY	AZ	BX	BY	BZ
537	7.75E5	50	116	129	44	90	45
538	7.75E5	50	63	50	44	89	134
547	7.74E5	50	115	50	45	89	135
548	7.74E5	50	64	129	45	90	44

The variation in the global stresses between the elements is nil. For the principal stresses the variation is between 4.83E5 and 4.84E5 which is less than 0.25% difference. A similar small variation can be seen for the von Mises stress which is not of interest to this problem and will not be discussed.

The most important stress for the concrete slab is the maximum principal tensile stress. Concrete is weak in tension with its tensile strength typically being about 10% of its compressive strength. Looking further at the stresses, it can be seen that the average maximum principal tensile stress is 4.84E5 N/m^2 or 0.484 MPa. Three of the four elements joining at node 744 give 0.484 MPa and the fourth gives 0.483 MPa.

It is usual for the stresses between adjacent elements to be compared. The stresses should be continuous and any discontinuity that is present is approximately the same as the error in the results. Graphs of the stress variation across elements should be plotted and the results should show a smooth variation, except where loads are applied or where dissimilar materials meet. For example Figure 8.12 shows the element and node numbers together with the maximum principal tensile stress along the base of the concrete slab, for $X = 1.5\text{--}2.5$ ($Y = 0.95$, $Z = 2.5$). Due to symmetry only values along $Z = 2.5$ need to be plotted. The values of the concrete tensile stress in the elements on either side of $Z = 2.5$ are plotted and confirm that the line $Z = 2.5$ is also a line of symmetry. Figure 8.13 shows the stress variation across the elements, between nodes 739 and 744 inclusive. At nodes 739, 740 and 744, the stress variation between adjacent elements is within computer round off error and can be considered as zero. Only at node 743, between elements 536 and 537 is the stress variation significant. The stress varies across the node between 0.366 MPa and 0.493 MPa, indicating an error in the stress of 0.127 MPa (0.493–0.366).

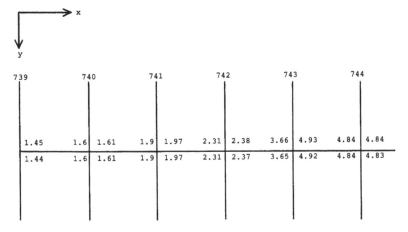

Figure 8.12 Sigma-1 principal tensile stress along base of concrete slab (stress values to be multiplied by $E + 5$ (N/m^2)).

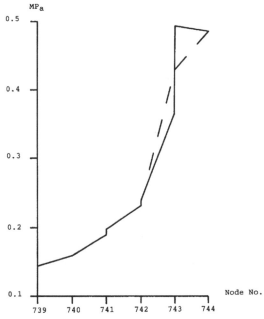

Figure 8.13 Stress change across nodes 739 to 744.

Two types of stress output can be obtained, averaged and unaveraged. This example has given unaveraged stresses for all nodes on an element. If averaged stress output had been required, then the stress values from the elements attached to a particular node would have been averaged, but as the elements joining each node have been made up of different layered materials, then stress averaging could not be performed unless requested.

The finite element mesh assumes a solid connection between all the different material layers (polystyrene/soil, soil/sand and sand/concrete). This would be correct if full friction developed between the layers and if nodal displacements were small. This is the assumption made in this chapter. If slippage were allowed between the layers, then additional data modules would have to be included, for example the additional module might describe a very thin interface layer. These additional modules are outside the scope of this chapter.

A further point to be noted by the PAFEC-FE user is that the results from one analysis may differ to those of a second analysis and thus all results should not be accepted without thought as the 'correct answer'. It is necessary to use engineering judgement in developing the model and interpreting the results. For example, the model used in this chapter started with 458 degrees of freedom. Mesh refinement was used and models were run with 1682, 2352, 3680 and 5285 degrees of freedom to provide asymptotic displacement and stress values. The mesh refinement results confirmed the answers given in this chapter.

8.6 Conclusions

In the PAFEC-FE analysis of the three-dimensional problem of a centrally loaded concrete slab on soil, element type 37100 has been used. In order to improve the accuracy of the results, a number of further analyses would be needed along the following lines:

(1) Use an element with one mid-side node 37110.
(2) Increase the number of elements, especially where the stresses are likely to vary rapidly. Use large elements in areas of small stress variation.
(3) The reactions at constraints should be calculated and compared with the applied loads.
(4) The shear and normal stresses at boundaries should be checked and should be equal to any applied loads.
(5) Model the area of the applied load as accurately as possible; e.g. a load spread over an area of say a tyre, should not be modelled as a point load.

Reference

D. Croney, *The Design and Performance of Road Pavements*, DoE, TRRL, HMSO, London (1977).

9 Static displacement and stress analysis

9.1 Introduction

Over the last few decades, design engineers involved in stress and displacement analysis problems have recognized the usefulness of the finite element method in solving static problems which challenged the conventional closed-form solutions.

This chapter deals with the static analysis of structures or components where the constitutive response of the materials involved is, or may be, approximated by isotropic elastic behaviour. Such a static analysis involves the calculation of stresses and displacements for situations where the following or a combination of the following are specified:

- Point loads applied at nodes
- Pressure uniformly distributed at the sides of elements
- Prescribed displacements at nodes

Static problems that come under the heading of framework analysis have already been dealt with in chapter 7. Furthermore, although axisymmetric analysis is examined in this chapter, it is restricted to solid bodies formed by revolution of a plane section. Similar analysis using thin shells, which were introduced in section 6.8, follows comparable lines.

9.2 Basic steps

In the development of a typical finite element computational model for static analysis, the following steps should be followed:

- Firstly it is necessary to select the appropriate mathematical model for the constitutive response of the materials involved. This implies that a decision must be taken as to whether it is acceptable to solve the problem by assuming an isotropic elastic response or whether anisotropy, plasticity or creep should be considered.
- For the purposes of this chapter we will assume isotropic elastic behaviour. A finite element discretization is then applied. The structure under consideration is sub-divided into a mesh of finite elements which might consist of two-dimensional (triangular and/or quadrilateral) or

three-dimensional (brick and/or wedge) elements. Furthermore, the appropriate boundary details are defined in terms of the specified static loading and constraint conditions that characterize the structure under consideration.

- After a successful computer run, the next step involves the presentation and interpretation of the static analysis results. The term successful here implies that no error messages have been generated by the finite element program and that any warning messages are considered acceptable by the PAFEC-FE user. Interpretation and checking of the computer output is enhanced significantly if the results are also presented graphically in the form of displaced shape, stress vector plots and stress contour plots.
- It has to be emphasized at this point that a successful computer run does not imply that the static analysis is correct. Having studied the finite element results, the PAFEC-FE user has also to consider the possibility that modifications or adjustments may be required at various stages in the finite element analysis. For example, it may be found that the finite element mesh used in the analysis is too coarse to capture the expected stress distribution and must therefore be refined. Alternatively unrealistic results may mean that a different and perhaps more accurate type of element must be used.

9.3   Considerations of symmetry

Economy of scale should always be the criterion that guides the steps followed during the finite element discretization of a structure, especially in three-dimensional static problems. It is all too easy to generate a mesh which defies the constraints imposed by local computing conditions. For this reason it is very important to explore any features of symmetry in the structure under consideration in order to reduce the analysis.

The following conditions must be satisfied, before a partial static analysis using the planes of symmetry is implemented:

- The geometry of the structure itself must be symmetric
- The boundary conditions (static loads and restraints) must also be symmetric.

If these two conditions are not fully attained then engineering judgement or intuition must be exercised by the PAFEC-FE user. For example, if the differences between two halves of a structure are sufficiently small for a fully symmetric structure to behave in essentially the same way as the actual structure, then a symmetric or anti-symmetric static analysis is possible.

If a structure has more than one plane of symmetry then the finite element mesh may be confined to a quarter or possibly only to an eighth of the complete structure, resulting in considerable economies. When only part of a structure is modelled then conditions of symmetry or anti-symmetry are applied to the cut plane, customarily in the form of boundary conditions or specified restraints.

There are of course circumstances where the analysed structure is characterized not by a symmetry with respect to a plane but by a symmetry with respect to one of the Cartesian axes. For such cases the solid structure may be generated when revolving by a full circle, a plane section around the axis of symmetry. Details for this specific type of static analysis are given in section 9.5.

9.4 Plane analysis

In considering physical structures, it is obvious that they are all inherently three-dimensional. However, in building mathematical models to represent the mechanical behaviour of such structures, there may be the option of effecting economies by characterizing the predominant behaviour as being two-dimensional. In the theory of elasticity there exists a special class of problems classified as plane problems, which can be solved more readily than the general three-dimensional problem since certain simplifying assumptions can be made.

When the development of the finite element analysis and the ramifications of its success to the engineering design of structures is examined, it is interesting to note that two-dimensional isotropic elastic static problems were the first successful examples to enhance the popularity of finite element analysis.

The geometry of a structure which permits its static analysis to be classified as a plane problem must satisfy the following:

- The structure should consist of a region of uniform thickness bounded by: (i) two parallel planes, customarily oriented normal to the z-axis and (ii) a closed lateral surface of arbitrary shape whose generators are parallel to the z-axis.
- Although the thickness of the structure, measured along the z-axis, must be uniform, it need not be limited. It may be very thick or very thin; in fact these two extremes represent the most desirable cases for plane static analysis, as will be pointed out later.

Readers not familiar with the applicable basic definitions of elasticity in static problems are referred to the elementary texts on the subject, in particular Mase (1970), the notation of which will be used for stresses and displacements in this chapter.

In employing two-dimensional finite element static analysis the stresses and displacements may be determined by using either the plane stress or the plane strain approach. These two important and different plane situations are normally distinguished on the basis of the geometrical characteristics of the structure under consideration. In general, the plane stress approach is employed when the structure is relatively thin in relation to its lateral dimension and the plane strain approach is used when the structure is very thick relative to its lateral dimensions.

9.4.1 Plane stress analysis

In many engineering problems it is often found that planes which are perpendicular to one of the reference axes are not subject to any stresses. Consider for example the analysis of stress at a free surface of a body, that is, a surface on which no forces are acting. Such a condition is known as the plane stress condition or the biaxial stress condition.

Since in plane stress conditions the structure thickness, measured along the z-direction, is small relative to its lateral dimensions, it is both permissible and advantageous to assume that stresses are constant over thickness. This implies that σ_{xx}, σ_{yy} and σ_{xy} do not vary throughout the thickness of such structures. In plane stress cases, the assumption is made that throughout the thickness of the structure, which may be typified by a thin plate subjected to loads applied in the xy plane, the stress components σ_{zz}, σ_{xz} and σ_{yz} are zero on both surfaces of the plate. The state of plane stress is then specified mathematically by making use of the stress components σ_{xx}, σ_{yy} and σ_{xy} which are functions of the x and y coordinates only. Furthermore, it is assumed that there is no constraint on the displacement field in the thickness or z-direction.

Taking the above assumptions into consideration as far as the stress and displacement components are concerned, the constitutive matrix for plane stress analysis may be expressed as follows (see section 6.5):

$$\frac{E}{1-v^2} \begin{bmatrix} 1 & v & 0 \\ v & 1 & 0 \\ 0 & 0 & \frac{1}{2}(1-v) \end{bmatrix}$$

When solving static two-dimensional problems the user of PAFEC-FE has to specify elements from the 36100 and/or the 36200 groups. There is no need to specify the type of analysis because plane stress is the default solution method.

9.4.2 Plane strain analysis

In the preceding section it was noted that the plane stress method is limited to situations where the structure thickness is small relative to lateral dimension. For those cases where the thickness or the z-dimension of a structure is very large, the condition clearly is the opposite extreme to plane stress. In particular, if z is large it may be assumed that the end sections of the structure are confined between two smooth and rigid plates so that the displacement along the z-direction is prevented. By symmetry there can be no displacement in the z-direction at the centre and, as it was made certain that there were no displacements in the same direction at the ends it is reasonable to expect that, throughout the length of the structure in the z-direction, there are no displacements. This condition, where strains can only occur in the xy plane, is known as the plane strain case.

A characteristic example that typifies the plane strain condition is the case of a long thick tube subjected to an external uniform pressure. If axial strain is prevented by the end conditions on the tube, then a plane strain analysis is required. For plane strain problems the thickness dimension along the z-axis is large compared with the typical dimensions in the xy plane and the structure is subjected to loads in the xy plane only. In such cases the stress perpendicular to the two-dimensional plane of interest (xy plane) is not zero and the in-plane displacements are considered as being independent of z.

The constitutive matrix for plane strain conditions is given by the following expression (see section 6.5):

$$\frac{E(1-v)}{(1+v)(1-2v)} \begin{bmatrix} 1 & v/(1-v) & 0 \\ v/(1-v) & 1 & 0 \\ 0 & 0 & (1-2v)/(2-2v) \end{bmatrix}$$

The above expression breaks down when Poisson's ratio reaches the value of 0.5, as the quantity $(1-2v)$ becomes zero and the expression outside the square brackets becomes infinite. This implies that the plane strain analysis constitutive matrix becomes inoperable for incompressible materials which by definition have $v = 0.5$. This is of little consequence in metals but in soils such as saturated clay (in its initial loading stages) or evaporites like rocksalt, is often assumed to have a Poisson's ratio equal to 0.5.

Clearly the assumption of $v = 0.5$ cannot be used in plane strain static problems. A simple way of side-stepping this stumbling block is to employ values of Poisson's ratio approximate to 0.5 but not equal to it. Although this solution appears to be simple and effective in dealing with the problem, Zienkiewicz (1971) warns that if this technique is adopted the approximation of solution deteriorates. An alternative procedure has been suggested and readers are referred for details to the work by Herrmann (1965), which involves the use of a variational formulation.

In certain situations of static analysis, especially in geotechnical applications, the plane strain conditions are not strictly valid even though the geometry of the structure satisfies fully the plane strain specifications. These are problems where the orientation of the axes of the principal initial stresses (i.e. geostatic stresses; see chapter 10 for definition) may not coincide with the plane of analysis. For such cases Zienkiewicz *et al.* (1978), has introduced the quasi-plane strain concept and shown its suitability in analysing a class of two-dimensional geotechnical problems. Adoption of this concept, allows extension of plane strain methods of stress analysis to the case of the stress distribution around long openings with any orientation in a triaxial stress field. Use of the quasi-plane strain analysis provides an analytical tool which is substantially cheaper than resorting to a fully three-dimensional analysis.

When dealing with plane strain analysis problems with PAFEC-FE, it is necessary to make use of the PLANE.STRAIN option in the CONTROL module, otherwise plane stress is assumed by default. The types of two-dimensional elements used for plane strain analysis are the same as those employed for plane stress analysis, namely the 36100 and 36200 groups.

9.5 Axisymmetric analysis

Although as already indicated, several static analysis problems are fully three-dimensional in that stresses and displacements in all three directions are important, very often the structure under consideration and the applied loading have an axis of symmetry. In such cases, a complete three-dimensional analysis may be considered as rather extravagant or indeed may not even be possible given local computing conditions. For such circumstances, an axisymmetric analysis should be used in which the finite element idealization models merely one generator plane employing two-dimensional elements.

The problem of stress distribution in bodies of revolution (axisymmetric solids) under axisymmetric loading is of considerable practical interest. Furthermore, even if the loading is not axisymmetric, provided that the structure is geometrically axisymmetric, it is possible to model the load by a number of Fourier components. For more details on this topic readers are encouraged to consult chapter 2.7.2 of the PAFEC Theory manual. The mathematical problems presented are very similar to those of plane stress and plane strain as, once again, the situation is two-dimensional.

When a three-dimensional solid structure is symmetrical about its centre-line axis and is subjected to loads that are symmetrical about this axis, the deformational behaviour is independent of the circumferential coordinate. Such a structure falls neatly into the typical category of axisymmetric analysis, where the constitutive matrix is given by the following expression (see PAFEC Theory manual):

$$\frac{E}{(1+v)(1-2v)} \begin{bmatrix} 1-v & v & 0 & v \\ v & 1-v & 0 & v \\ 0 & 0 & \frac{1}{2}-v & 0 \\ v & v & 0 & 1-v \end{bmatrix}$$

Examination of the above constitutive matrix indicates that, as in the case of plane strain analysis, the expression becomes inoperable for $v = 0.5$, in which case the reader should refer to the discussion presented in the previous section.

The use of axisymmetric analysis employing PAFEC-FE requires that the axis of symmetry is always the x-axis with y being the radial coordinate, in which case the hoop strain is calculated as the product of the radial displacement with the term $(1/y)$. However, in the special case where a structure lies partly on the axis of symmetry, it is never necessary to evaluate $(1/y)$ on the axis of symmetry where $y = 0$, since PAFEC-FE is using the Gaussian integration method. When element stresses are required on the x-axis, PAFEC-FE calculates the hoop strain by setting it equal to the radial strain. For this reason, care has to be exercised in the interpretation of results at the points that are located on the axis of symmetry.

For the mesh idealization of the generator plane used in axisymmetric analysis, PAFEC-FE employs the ordinary two-dimensional elements from the 36100 and 36200 groups that are normally used for plane stress or plane strain analysis. Consequently, analysis of axisymmetric structures with PAFEC-FE requires the use of the AXISYMMETRIC option in the CONTROL module, otherwise plane stress will be assumed by default.

9.6 Three-dimensional analysis

As was demonstrated in the previous two sections, depending on the geometry and loading, structures under investigation may be represented as plane stress or plane strain configurations or axisymmetric solids. However, the most accurate representation is undoubtedly the three-dimensional solid.

However, it must be borne in mind that three-dimensional elements are expensive to use and should only be employed when stresses vary in three dimensions. Consequently, although use of three-dimensional elements clearly satisfies all practical cases in static analysis, users of PAFEC-FE should always investigate whether the various two-dimensional approximations offer an adequate model.

9.7   Static loading

In static analysis, if it is required to apply point loading to a structure, it is merely necessary to establish the direction of the required load at the relevant node and make the appropriate entry in the LOADS module. If on the other hand, there is a need to model a distributed load acting on an element, the situation becomes slightly more involved. In these cases, it is suggested that users should refer to Tables 2.12, 2.13 and 2.14 in the PAFEC Theory manual. PAFEC-FE has a particular approach to specifying the equivalent nodal loads that correspond to the various types of distributed loading conditions.

For static analysis where there is a need to apply a constant or varying pressure load onto a face of a structure, the PAFEC Theory manual distinguishes two cases:

- The structure is modelled in full three-dimensional idealization employing three-dimensional elements. Here the equivalent nodal loads are generated through a complicated code whereby the magnitude of the pressure is determined by interpolating between the nodes that form the vertices of the surface where pressure is exerted. In general the equivalent loads involve three components at each of the element nodes on the surface in question and the pressure always acts inwards on the elements.
- The structure is modelled by using two-dimensional elements in plane stress, plane strain or axisymmetric configuration. Here the integration to obtain the equivalent nodal loads is quite straightforward and once more the pressure acts inwards on the elements.

PAFEC-FE offers three different modules to specify pressure loading, namely PRESSURE, SURFACE.FOR.PRESSURE and FACE.LOADING. When using the PRESSURE module, since by default any node not defined in the data will be assumed to have zero pressure, users must ensure that element mid-side nodes are included in the LIST.OF.NODES or in the range of nodes defined by the START, FINISH and STEP.

9.8   Restraints

Boundary restraint conditions are generally expressed in a straightforward manner. In very large problems, the user of PAFEC-FE may have to decide on the extent of the structure which should be taken into account and the boundary restraint conditions which should then be applied.

In problems involving symmetrical configurations, care should be taken at any point which is on more than one plane of symmetry to ensure that all the constraints required for each of the planes are applied.

When declaring the applied restraints by specifying the PLANE in the RESTRAINTS module, for situations where the plane is a curved surface (i.e. cylindrical coordinates), care must be taken in ensuring that PAFEC-FE locates all the points lying on the curved surface. It may be necessary to reduce the TOLERANCE in the CONTROL module to ensure that all the nodes will be included in the PLANE specification.

9.9 Examples of static analysis using PAFEC-FE

In the following examples use is made of the family of two-dimensional and three-dimensional isoparametric elements. The most popular two-dimensional element is the eight noded isoparametric quadrilateral. In three dimensions its equivalent is the 20-noded hexahedral or brick element, an element with eight nodal corners and six quadrilateral sides and three degrees of freedom at each node.

The term isoparametric implies equal (iso-) parametric description of the unknown displacement and geometry of the element. The basic idea is to express both the displacement and the geometry of the element by using the same interpolation functions.

The concept of isoparametric elements has been commonly used for finite element formulations. It offers a number of advantages, including efficient integrations and differentiations and accurate modelling of curved and arbitrary geometrical shapes. A detailed description of these elements is given in the PAFEC Theory manual, Chapter 2.4.

Three examples all referring to an internally pressurized thick walled cylinder, are analysed using equivalent two-dimensional, axisymmetric and fully three-dimensional analysis. The cylinder is mild steel, with an internal diameter of 100 mm, an external diameter of 180 mm and is subjected to an internal pressure of 70 MPa.

The results are compared with available (Timoshenko and Goodier, 1970) closed-form solutions and as can be seen from the following tables and graphs the output from PAFEC-FE is exceptionally good for all three analyses. Furthermore, it is interesting to note that the two-dimensional idealization of the problem produced nearly identical results to the three-dimensional analysis and the axisymmetric modelling (see Tables 9.1 and 9.2).

　　　　　　　ENGINEERING ANALYSIS USING PAFEC

Table 9.1 Comparison of theoretical solution with PAFEC-FE: solution for stress results

Radial distance (mm)	Theory	Analysis 3 (3D)			Analysis 1 (2D)	Analysis 2 (axisym.)
		Top	Middle	Bottom		
Tangential stress (MPa)						
50.00	132.50	133.00	133.00	133.00	133.00	133.00
53.33	120.24	120.00	120.00	120.00	120.00	120.00
56.67	110.08	110.50	110.50	110.50	111.00	110.00
60.00	101.56	101.00	101.00	101.00	101.00	101.00
63.33	94.36	94.60	94.62	94.65	94.60	94.60
66.67	88.20	88.10	88.10	88.10	88.10	88.10
70.00	82.91	83.05	83.15	83.10	83.10	83.10
73.33	78.32	78.20	78.30	78.30	78.20	78.20
76.67	74.31	74.45	74.45	74.45	74.40	74.40
80.00	70.80	70.70	70.80	70.80	70.70	70.70
83.33	67.70	67.75	67.77	67.80	67.80	67.80
86.67	64.95	64.90	64.90	64.90	64.90	64.90
90.00	62.50	62.60	62.60	62.60	62.50	62.60
Radial stress (MPa)						
50.00	−70.00	−68.80	−68.90	−68.80	−68.80	−68.70
53.33	−57.74	−58.30	−58.40	−58.30	−58.30	−58.30
56.67	−47.58	−46.57	−46.55	−46.55	−46.60	−46.60
60.00	−39.06	−39.40	−39.40	−39.40	−39.40	−39.40
63.33	−31.86	−31.20	−31.20	−31.25	−31.20	−31.20
66.67	−25.70	−25.90	−25.90	−25.90	−26.00	−26.00
70.00	−20.41	−20.00	−19.95	−20.00	−20.00	−20.00
73.33	−15.82	−16.00	−16.00	−16.00	−16.00	−16.00
76.67	−11.81	−11.50	−11.50	−11.50	−11.50	−11.50
80.00	−8.30	−8.42	−8.39	−8.41	−8.42	−8.42
83.33	−5.20	−5.04	−4.97	−4.98	−4.99	−4.99
86.67	−2.45	−2.55	−2.53	−2.54	−2.54	−2.54
90.00	0.00	0.09	0.11	0.08	0.19	0.07

The data together with the associated graphical output are given below for the three analyses.

9.9.1　Analysis 1

Here a plane strain analysis is undertaken for the cross-section of the thick walled cylinder. Because of the radial symmetry of the problem there is no need to analyse the complete cross-section. Here half of the cross-section has been modelled (see Figures 9.1 and 9.2), although a quarter of the cross-section with appropriate restraints would have been adequate. It was decided to analyse the half cross-section so that an important point concerning PAFEC-FE's idiosyncratic treatment of restraints might be highlighted.

Table 9.2 Comparison of theoretical solution with PAFEC-FE: solution for displacement results

Radial distance (mm)	Radial displacement (μm)				
	Theory		Analysis 3 (3D)	Analysis 1 (2D)	Analysis 2 (axisym.)
	Plane stress	Plane strain			
50.00	36.72	35.38	36.71	35.40	36.72
53.33	35.10	33.67	35.09	33.69	35.10
56.67	33.72	32.19	33.71	32.22	33.71
60.00	32.52	30.91	32.51	30.93	32.52
63.33	31.49	29.78	31.48	29.81	31.49
66.67	30.59	28.80	30.59	28.83	30.59
70.00	29.82	27.93	29.81	27.96	29.82
73.33	29.15	27.17	29.14	27.20	29.14
76.67	28.56	26.50	28.55	26.53	28.56
80.00	28.05	25.90	28.05	25.93	28.05
83.33	27.62	25.37	27.61	25.40	27.61
86.67	27.24	24.91	27.23	24.93	27.24
90.00	26.91	24.49	26.91	24.52	26.91

Since the structure is symmetric, the two sides of the model lying on the plane of symmetry of the cross-section must be restrained from moving along the z-direction (see Figure 9.2). This can be achieved by introducing appropriate restraints to nodes 1 and 5. Furthermore, it is important to constrain node 6, or equally well node 2, to prevent the structure from sliding freely along the y-direction. If this step is not taken, the problem is not uniquely determined and PAFEC-FE automatically imposes an additional restraint. The outcome is unpredictable since PAFEC-FE chooses to restrain the highest numbered of the generated nodes.

Of course, if a quarter of the cross-section was to be modelled, then it would have been adequate to restrain each side from moving in the direction normal to itself. There would be no need to constrain an additional point in order to determine a unique solution.

We note in the following data that it was necessary to reduce the tolerance so that the SURFACE.FOR.PRESSURE module could locate all the nodes lying on the inner surface of the steel cylinder. The Young's modulus and the pressure are specified in MPa while the density is expressed in kg/mm^3, because the dimensions of the cylinder are given in millimetres.

The calculated radial displacements from this plane strain analysis are compared with the theoretical predictions (see Figure 9.3 and Table 9.2) and as is observed, once more the correlation is excellent. In Figure 9.2 the deformed shape of the half cross-section is shown where it can be seen that node 6 did not suffer any displacement as specified. The stress vectors corresponding to the distribution of radial and tangential stresses are shown in Figure 9.4, while Figures 9.5 and 9.6 show the minimum and maximum stress distribution respectively.

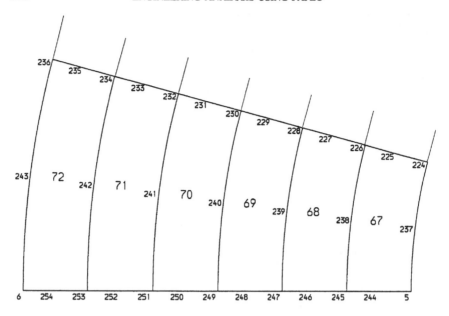

Figure 9.1 Part of the finite element mesh used for two-dimensional analysis of a thick wall cylinder (node numbers and element numbers are shown).

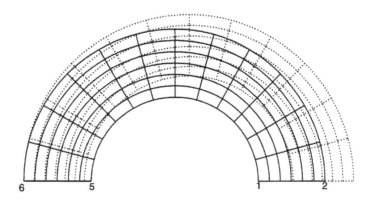

Figure 9.2 Deformed shape of a thick wall cylinder section subjected to internal pressure (two-dimensional solution).

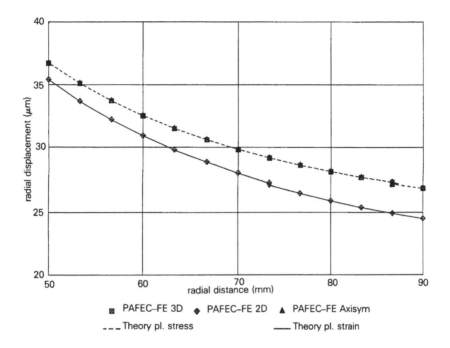

Figure 9.3 Distribution of radial displacement for a thick wall cylinder (comparison of theoretical solution with PAFEC-FE results).

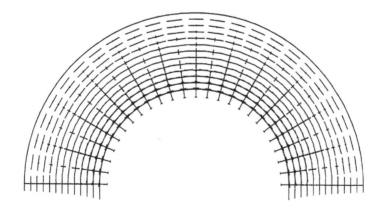

Figure 9.4 In-plane stress vectors for two-dimensional analysis of an internally pressurized thick wall cylinder.

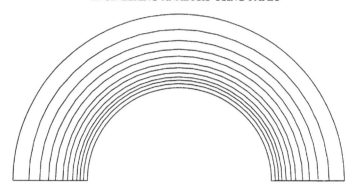

Figure 9.5 Minimum (most negative) principal in-plane stress contours for two-dimensional analysis of an internally pressurized thick wall cylinder.

Figure 9.6 Maximum (most positive) principal in-plane stress contours for two-dimensional analysis of an internally pressurized thick wall cylinder.

```
TITLE Analysis 1 — Static analysis, 2-D idealization of internally
* pressurized cylinder
C
CONTROL
PLANE.STRAIN
TOLERANCE = 10E–2
CONTROL.END
C
AXES
AXISNO   RELAXISNO   TYPE   NODE.NO   ANG1
   4         1          1       7       –90
C
NODES
NODE.NUMBER   AXIS.NUMBER   X   Y   Z
     1             2         0  50   0
     2             2         0  90   0
```

3	2	0	50	90
4	2	0	90	90
5	4	50	0	0
6	4	90	0	0
7	1	0	0	0

C
MATERIAL

MATERIAL.NUMBER	E	NU	RO
1	209E3	0.3	7.8E–6

C
PLATE.AND.SHELLS

PLATE.OR.SHELL.NUMBER	MATERIAL.NUMBER
1	1

C
MESH

REFERENCE	SPACING.LIST
12	12
6	6

C
PAFBLOCKS

TYPE	ELEMENT.TYPE	PROPERTIES	N1	N2	TOPOLOGY
1	36210	1	6	12	1 2 5 6 0 3 4

C
RESTRAINTS

NODE.NUMBER	PLANE	AXIS.NUMBER	DIRECTION
1	3	2	3
5	3	2	3
6	0	2	2

C
SURFACE.FOR.PRESSURE

PRESSURE.VALUE	NODE	PLANE	AXIS
70	1	2	2

C
C
IN.DRAW

TYPE.NUMBER	INFORMATION.NUMBER
2	23

C
C
OUT.DRAW

PLOT.TYPE
  1
 20
 30
 31
C
C
END.OF.DATA

9.9.2 Analysis 2

Here the same structure is analysed again, this time using an axisymmetric analysis that is achieved by revolving the mesh idealization shown in Figures 9.7 and 9.8.

The analysis is essentially a three-dimensional one corresponding to plane stress conditions because we have restrained the cylinder only from one side, the side containing node 1 (see Figures 9.7 and 9.8 and RESTRAINTS module in the following set of data).

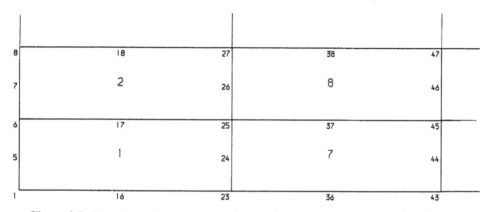

Figure 9.7 Part of the finite element mesh used for axisymmetric analysis of a thick wall cylinder (node numbers and element numbers are shown).

Figure 9.8 Deformed shape of a thick wall cylinder section when subjected to internal pressure (axisymmetric solution).

Comparison of the axisymmetric analysis results with the three-dimensional analysis and the analytical solution is very impressive as one can see from Table 9.1 (for stresses) and Table 9.2 (for radial displacement).

The deformed shape obtained from the OUTDRAW module (Figure 9.8) indicates a perfect plane stress condition where the free end of the cylinder under the influence of the internal pressure contracts.

Figure 9.9, which is also obtained as output of the OUTDRAW module, shows the perfect orientation that the two stresses have, this being parallel (for the axial stress) and normal (for the radial stress) to the axis of the cylinder.

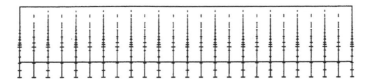

Figure 9.9   In-plane stress vectors for axisymmetric analysis of an internally pressurized thick wall cylinder.

Figure 9.10   Minimum (most negative) principal in-plane stress contours for axisymmetric analysis of an internally pressurized thick wall cylinder.

Figure 9.11   Maximum (most positive) principal in-plane stress contours for axisymmetric analysis of an internally pressurized thick wall cylinder.

Figure 9.12   Hoop (tangential) stress contours for axisymmetric analysis of an internally pressurized thick wall cylinder.

Figures 9.10, 9.11 and 9.12 requested from the OUTDRAW module, show the contour drawings for the minimum, maximum and hoop (tangential) stress distribution.

```
TITLE Analysis 2 — Static analysis, axisymmetric idealization of
* internally pressurized cylinder
C
CONTROL
AXISYMMETRIC
CONTROL.END
C
NODES
```

```
NODE.NUMBER  X    Y
   1              0   50
   2              0   90
   3            240   50
   4            240   90
C
MATERIAL
MATERIAL.NUMBER  E       NU   RO
   11               209E3  0.3  7.8E-6
C
PLATE.AND.SHELLS
PLATE.OR.SHELL.NUMBER   MATERIAL.NUMBER
   1                          11
C
MESH
REFERENCE   SPACING.LIST
   6            6
   12           12
C
PAFBLOCKS
TYPE   ELEMENT.TYPE   PROPERTIES   N1  N2  TOPOLOGY
   1     36210            1          6   12  1  2  3  4
C
RESTRAINTS
NODE.NUMBER   PLANE   DIRECTION
   1            1        1
C
SURFACE.FOR.PRESSURE
PRESSURE.VALUE   NODE   PLANE
   70             1       2
C
IN.DRAW
TYPE.NUMBER   INFORMATION.NUMBER
   2              23
C
OUT.DRAW
PLOT.TYPE
   1
   20
   30
   31
   33
C
END.OF.DATA
```

9.9.3 *Analysis 3*

In computing terms this has been the most expensive analysis, yet the results are not more accurate. This analysis requires approximately 40 times the working storage of the comparable analyses used in the two-dimensional and axisymmetric cases. In terms of execution time, depending on the particular operating system, there may be an increase by a factor of 20. This illustrates that three-dimensional analysis is not always necessary to achieve good results.

Only a quarter of the cylinder has been modelled (see Figure 9.13) and as was mentioned earlier in the discussion of the two-dimensional analysis, it was only necessary to restrain the nodes lying on the planes of quarter symmetry from moving normal to these planes of symmetry. In addition to the constraints at the planes of symmetry, the lower end of the cylinder was prevented from moving along the axis.

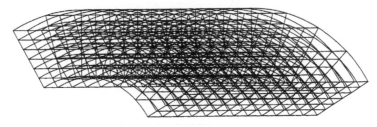

Figure 9.13 Finite element mesh representing a quarter of a thick wall cylinder, as used in the three-dimensional analysis.

Comparison of the three-dimensional results both in terms of stresses (Table 9.1) and in terms of the radial displacement shows remarkable agreement.

Furthermore, since this three-dimensional solution is essentially a plane stress analysis (see Figure 9.14 depicting the deformed shape of the pressurized cylinder), the radial displacements are the same with the axisymmetric analysis but slightly higher than the plane strain analysis (see Figure 9.2). On the other hand, if the ends of the cylinder were restrained from moving along the longitudinal axis, the three-dimensional solution would have simulated plane strain conditions.

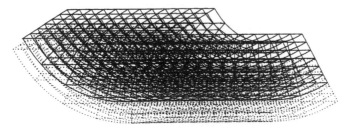

Figure 9.14 Deformed shape of thick wall cylinder subjected to internal pressure (three-dimensional analysis of quarter of the structure).

The stress distribution in the form of stress vectors for the edge of the cylinder is shown in Figure 9.15, while comparison of the theoretical and the three-dimensional finite element stress distribution (for radial and tangential components) show a perfect fit in Figure 9.16.

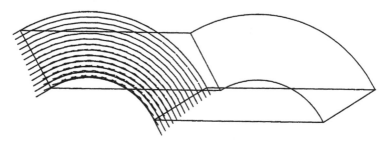

Figure 9.15 In-plane stress vectors at the end of an internally pressurized thick wall cylinder (three-dimensional analysis).

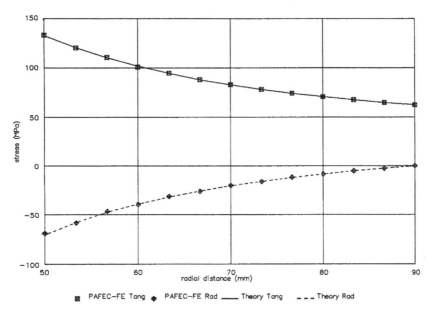

■ PAFEC–FE Tang ◆ PAFEC–FE Rad ⎯⎯ Theory Tang ⎯ ⎯ Theory Rad

Figure 9.16 Distribution of radial and tangential stress components for a thick wall cylinder (comparison of theoretical solution with PAFEC-FE results).

Finally stress contours for the maximum principal stress distribution for the edge of the cylinder are shown in Figure 9.17 as requested by the SECTION.FOR.PLOTTING module.

TITLE Analysis 3 — Static analysis, 3-D idealization of internally
* pressurized cylinder
C

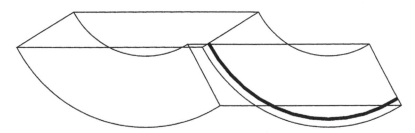

Figure 9.17 Maximum (most positive) principal in-plane stress contours at the end of an internally pressurized thick wall cylinder (three-dimensional analysis).

```
CONTROL
TOLERANCE = 10E-2
CONTROL.END
C
NODES
AXIS.NUMBER = 2
NODE.NUMBER     X    Y    Z
     1          0   50    0
     2          0   90    0
     3          0   50   45
     4          0   90   45
     5          0   50   90
     6          0   90   90
     7        280   50    0
     8        280   90    0
     9        280   50   45
    10        280   90   45
    11        280   50   90
    12        280   90   90
    13        200   80   30
C
MATERIAL
MATERIAL.NUMBER  E      NU  RO
     1         209E3   0.3  7.8E-6
C
MESH
REFERENCE   SPACING.LIST
    6           6
   12          12
C
PAFBLOCKS
TYPE   ELEMENT.TYPE  PROPERTIES  N1  N2  N5  TOPOLOGY
  1      37110           1        6  12   6     1  2 5 6 7 8
  *                                          11 12 0 3 4 0 0
  *                                           0  0 0 0 9 10
```

```
C
RESTRAINTS
NODE.NUMBER   PLANE   AXIS.NUMBER   DIRECTION
     1           1          1             1
     1           3          1             3
     5           2          1             2
C
SURFACE.FOR.PRESSURE
PRESSURE.VALUE   NODE   PLANE   AXIS
     70            1      2       2
C
IN.DRAW
TYPE.NUMBER   INFORMATION.NUMBER   NODE.NUMBER
     2               0                 13
C
OUT.DRAW
NODE.NUMBER = 13
DRAWING.NUMBER   PLOT.TYPE
     1               1
     2              20
     3              30
C
SECTION.FOR.PLOTTING
DRAWING.NUMBER   PLANE   AXIS.NUMBER   N1
     2             1         2          1
     3             1         2          1
C
END.OF.DATA
```

References

L.R. Herrmann, Elasticity equations for incompressible, or nearly incompressible materials by the variational theorem, *J.A.I.A.A.* **3** 1896 (1965).

E.G. Mase, *Theory and Problems of Continuum Mechanics*, Shaum's Outline Series, McGraw-Hill, New York (1970) p. 221.

S.P. Timoshenko and J.N. Goodier, *Theory of Elasticity*, 3rd edn., McGraw-Hill, New York (1970) p.567.

O.C. Zienkiewicz, *The Finite Element Method in Engineering Science*, McGraw-Hill, London (1971) p.521.

O.C. Zienkiewicz, R.L. Taylor and G.N. Pande, Quasi-plane strain in the analysis of geological problems, in *Computer Methods in Tunnel Design*, The Institution of Civil Engineers, London (1978) pp.19–40.

10 Geotechnical engineering applications

10.1 Introduction

Rational geotechnical engineering practice requires competent techniques for predicting ground response to excavation activities. In particular, there is a need for numerical methods, such as finite element analysis, which allow parametric studies to be pursued quickly and efficiently, so that a number of operationally feasible construction options can be evaluated for their geotechnical rationality. In employing the finite element method, the objective is to assess the stability of the ground by identifying the stress and displacement distribution around excavations or ground structures.

The principle of using the finite element method in the study of geotechnical problems, has been established for some while and the extensive literature on the subject indicates that geotechnical engineers have had a great deal of success in using this type of analysis (Burt, 1978).

In a finite element analysis, the initial ground stress field and the deformation and strength characteristics assumed for the geological materials involved, will strongly affect the computational procedures. Unfortunately, the exact nature of the initial ground stresses and the experimental determination of mechanical properties of the various geological materials is seldom satisfactory, and so an idealized model is used. In this respect, finite element analysis is an excellent design tool, since it allows the engineer to repeat the analysis using a combination of ground stress conditions with a series of values for the material constants, spanning over the range of the experimental uncertainty.

On many occasions finite element investigations treat geological materials, at least as a first approximation, as an isotropic linear elastic material. However, for relatively high stresses when yield occurs, geological materials depart from elastic behaviour and begin to suffer permanent deformation, whereby an elasto-plastic analysis may be adopted. Another departure from linear elasticity exists for materials that creep, such as evaporites and chalks, in which case finite element analysis is implemented by a series of incremental solutions over sufficiently small time steps.

As a result, finite element packages aiming specifically at the geotechnical engineering field, are expected to be able to deal with the following conditions:

- Geological materials that creep (i.e. are characterized by a time-dependent mechanical response), behave in an elasto-plastic manner, or have an anisotropic character (as a result of their bedded or layered nature)
- Joints or other geological discontinuities in rock formations
- The application of gravity loading and implementation of ground failure criteria
- The simulation of the construction sequence of a ground excavation by modelling satisfactorily both the removal of the ground and the action of ground support systems

PAFEC-FE can satisfy all the aforementioned requirements, with the exception of the modelling of joints or discontinuities in rocks. However, work is under way to implement in PAFEC-FE a joint element type, similar to the one introduced by Goodman (1976), capable of simulating the action of discontinuities in rock masses. In the meantime, modelling of geological discontinuities resulting in bed separation, may be achieved using the GAPS module (Reed, 1988).

The finite element method has the advantage of being a very flexible technique. Once the geometry of a particular problem has been defined, the loads imposed on the model and the properties attributed to the associated geological materials can be varied with ease. This is very useful in geotechnical problems in general, since it becomes a very simple task to study the sensitivity of the stability of ground structures to changes in those properties where the major uncertainty exists such as the material parameters of ground and the initial ground stresses.

The conflict in finite element analysis associated with geotechnical problems is the requirement to obtain a reasonable accuracy whilst keeping the computing requirements within the bounds laid down by the local operating system. It is all too easy to specify a model which may necessitate several hours of computing time for analysis and may require massive storage for intermediate calculations. This factor is more apparent in the study of near surface and underground excavations where the engineer has to model the geological medium surrounding the structure, whereas in other engineering applications, the engineer is simply expected to model the structure itself. Consequently, it is an easy trap for the inexperienced user of PAFEC-FE, especially if PAFBLOCKS are used, to generate a model with many thousands of nodes.

This is even more obvious when working in three dimensions, where the number of nodes is increased approximately 2.5 times by the introduction of the third dimension of similar proportion. To avoid such pitfalls, the engineer has to design an optimum mesh which not only will offer the greatest possible degree of detail around the excavation, but will also reduce the number of nodes away from the opening further into the ground mass where detailed accuracy may not be required.

Of the many modules available in PAFEC-FE the following are typically utilized in problems associated with the assessment of stability of ground structures:

- NODES for specification of nodal coordinates.
- PAFBLOCKS for automatic mesh generation and ELEMENTS to define single elements.
- AXES to reference nodes in local coordinates pertinent to the geometry of the ground excavation under consideration.
- RESTRAINTS to restrain the unloaded edges of the model located below the excavation level (and occasionally in the lateral direction), and to introduce planes of symmetry in simplified geometries.
- MATERIAL to specify the mechanical characteristics of the geological materials involved in the analysis and PLATES.AND.SHELLS in a two-dimensional approach.
- STRESS.ELEMENT to restrict the stress determination to the elements of interest.
- EXTERNAL.FORCES to request the calculation of forces acting on specific nodes as a result of the application of prescribed loading conditions.
- LOCAL.DIRECTIONS to specify an axis related to the element face that forms part of the excavation periphery, in order to input radial loads which have been calculated using the EXTERNAL.FORCES module
- LOADS and PRESSURE (or SURFACE.FOR.PRESSURE or FACE.LOADING) to apply node loads and uniformly distributed pressure, respectively, over boundaries.
- IN.DRAW and OUT.DRAW to request suitable graphic output and GRAPH to obtain stress or displacement plots along specific lines of interest, normally emanating in a radial direction from the walls of a ground excavation. Furthermore the module SECTION.FOR.-PLOTTING may be used to restrict the OUT.DRAW output only to specific areas of interest, normally close to the walls of a ground excavation.
- CREEP.LAW to specify the time dependent constitutive relationship of the geological materials that creep, by using the empirical power law

$$\varepsilon = A \, \sigma^n \, t^m$$

- PLASTIC.MATERIAL to specify the non-elastic stress–strain relationship for geologic materials that exhibit elasto-plastic behaviour.
- YIELDING.ELEMENTS to specify the elements that have either creep properties or plastic properties.

- FAILURE.CRITERIA to define the strength parameters of the geological materials involved by assuming an idealized Coulomb–Navier failure criterion. In its simplest form this module models the ground as a 'non-tension' material (Zienkiewicz, 1977). If required, the PAFEC-FE source code may be modified by means of the USE. control option, to define other suitable failure criteria (see section 3.5.4).
- GRAVITY to simulate the effect of the vertical gravitational stress field that is normally imposed on the geological formations surrounding a ground structure.
- INCREMENTAL used with plasticity analysis to define the magnitude of the load increments to be applied as a percentage of the total load. Also used with creep analysis to define the time steps used during the non-linear numerical procedure.
- LAMINATES to define the relevant geometric details (such as curvature of element, thickness, etc.) when PAFEC-FE is used to model stratified, bedded or laminated geological materials. For such materials the module ORTHOTROPIC.MATERIAL is also used to specify the material parameters (i.e. compliances in the principal directions, cross-compliances, shear compliances, etc.) describing the orthotropic response.
- POLAR.DIRECTIONS a very useful module that allows radial and tangential degrees of freedom directions to be assigned to the surface of excavations that have cylindrical or spherical shape (i.e. tunnels or mine shafts with circular cross-sections). Use of this module facilitates the interpretation of the displacement results, since they are produced along the polar tangential and radial directions instead of along the Cartesian x- and y-axes.

In using PAFEC-FE to assess the stress and displacement distribution around ground structures, there are a number of problems that the engineer is expected to consider:

- Application of the specified initial ground stress field
- Simulation of the process of the excavation of the ground opening
- Positioning of the boundary of the finite element mesh in relation to the walls of the ground excavation
- Modelling the presence of underground support systems

In the following sections, the aforementioned problems are addressed and suitable approaches are suggested.

10.2 Initial ground stress field

In conventional finite element analysis, the geometry of a structure and its operating duty define the loads imposed on the system. In a geotechnical problem, the geological medium is subjected to initial ground stress prior to excavation, also known as geostatic stress. The final, post-excavation state of stress in the ground structure is the resultant of the initial geostatic stress and the stress concentrations induced by the creation of the opening.

The incorporation of the initial geostatic stresses in a finite element analysis may take place by applying gravity loading to the unstressed ground model or may be specified directly by appropriate pre-stressing of the elements modelling the ground mass.

To simulate the existence of the geostatic stress field by employing the gravity technique, the mesh that models a particular geometry corresponding to a ground excavation may be loaded by any of the three following methods:

(i) The top surface of the mesh is loaded by applying a uniformly distributed pressure equivalent to the vertical static load, exerted by the weight of the overlaying strata, by using the PRESSURE (or SURFACE.FOR.PRESSURE or FACE.LOADING) module. If on the other hand the finite element mesh extends up to the ground surface, then the top surface of the mesh is left unloaded.

(ii) The GRAVITY module is used to increment the stress according to the depth of the model below the loaded surface.

(iii) The horizontal geostatic stress may be introduced by:

- Either the application of constant pressure to the side boundaries of the mesh, using one of the pressure modules. The pressure is set equal to the value of the pressure applied to the top surface of the mesh and multiplied by the appropriate k_o value (= ratio of horizontal to vertical geostatic stress).

- Or by using RESTRAINTS to ensure that the side boundaries of the mesh are free to move only in the vertical direction, thus introducing horizontal stresses as a result of the lateral confinement. In this case k_o will be a function of the Poisson's ratio of the geological materials involved and as such will automatically be determined by the finite element analysis.

At points located above and below an underground opening, the actual horizontal stress component is expected to vary linearly with depth below the surface. However, in the case of tunnels where opening size is often relatively small in comparison with depth, it is acceptable to assume that the horizontal geostatic stress component may be modelled as a constant pressure loading.

Alternatively the introduction of the geostatic stress field may be achieved by the method of direct stress specification (Naylor et al., 1981). Such an approach also caters for effective stress analysis by allowing for the presence of initial pore water pressure. Although this technique is successful in providing a satisfactory solution to the simulation of geostatic stresses, its implementation in PAFEC-FE is not as easily achieved as the gravity method which is the recommended method when using PAFEC-FE.

10.3   Excavation simulation

Deep open-air excavations are normally developed in stages, gradually progressing deeper in excavation steps. Furthermore, large openings, typical of those used in underground power station machine halls, are usually excavated in a series of multiple faces or as a top face advance followed by one or more benches to remove the bulk of the cavern. Consequently, to model realistically such excavation processes, the numerical technique employed must be capable of simulating an arbitrary excavation sequence.

A detailed description of how to implement an excavation process, whether it takes place in a single stage or a multi-stage, if the geostatic stresses are to be defined by the method of direct stress specification is given by Naylor *et al.* (1981).

Furthermore, the simulation of an excavation process using finite element analysis, has also been described by Kulhawy (1974) and later by Breckles (1978) and Breckles *et al.* (1981), and presents three possible methods:

- Gravity difference method
- Relaxation approach
- Stress reversal technique

In the following sections these techniques are examined in detail.

10.3.1   Gravity difference method

This method, also called the 'gravity turn-on' approach, requires that two consecutive analyses are conducted. The first analysis loads the block of ground without the excavation, and the second applies gravity load to the same block of ground, but with the excavated ground removed, or sometimes with the excavated elements given a very low modulus value. The difference between the two analyses produces the stresses and displacements that correspond to the creation of the ground excavation.

The gravity difference approach is limited since it cannot distinguish between a multi-stage excavation process and an excavation that takes place in one phase. This is because in the gravity difference method the resulting stresses are independent of the construction sequence (which may include several excavations) only if the geological material satisfies the requirement of homogeneity, isotropy and, most importantly, linear elasticity.

As a result this approach is considered as rather unsuitable in simulating the exact sequence of ground excavation processes and is therefore limited to the case of elastic ground and single phase excavation.

10.3.2 Relaxation approach

This method is also known as the residual stress approach. The final geometry of the underground opening is established in the finite element mesh. The model is loaded using the prescribed geostatic stress field, while the excavation face is subjected to some arbitrary stress level corresponding to the support pressure previously provided by the ground inside the excavation periphery. The excavation face is then relaxed by stepwise unloading until the desired values of stress or displacement are obtained in the elements that form the excavation periphery.

With this approach, it is difficult to simulate an excavation sequence, the stress in the elements removed during the excavation is not considered, and the relaxation is controlled solely by the elements which form the rock mass surrounding the excavation.

A possible consequence of following the relaxation approach for geological materials that behave in a time-independent but non-elastic manner, is the underestimation of the displacements and the overestimation of the stresses. This is to be expected since the relaxation approach is not considering the stresses in the elements removed during excavation and consequently is not modelling the stress redistribution that takes place as a result of the elasto-plastic response of the ground.

10.3.3 Stress reversal technique

The excavation is simulated by first evaluating the stress that would exist along the potential excavation surface as a result of applying the geostatic loading to a suitable mesh. The ground excavation is in a 'solid' or unexcavated state at this stage.

The stresses along the proposed excavation surface are then calculated in terms of the equivalent forces at each node. The signs of the forces are reversed, and the radial component is then applied to the nodes in a second mesh, in which the excavation step has been simulated. By combining the before and after analyses, the stresses and displacements may be determined.

Taking into consideration the disadvantages of the other two methods, it is suggested that the stress reversal method is best suited to simulate numerically the process of the excavation of ground excavations. Work by Breckles et al. (1981), where PAFEC-FE was used, presents a typical application of the stress reversal technique in simulating the excavation of underground openings.

When using the stress reversal technique, the EXTERNAL.FORCES module is ideally suited to the simulation of the excavation process, since it eliminates the need manually to re-evaluate element stresses into equivalent forces for application to the excavated boundary. Furthermore by making use of the LOCAL.DIRECTIONS module, one may obtain the forces calculated by PAFEC-FE oriented in terms of radial and tangential components. In this case all that is required is to reverse the sign of the radial force and apply it to the nodes of the opening's periphery.

10.4 Boundary proximity

Use of finite element analysis for the assessment of the stability of ground excavations, requires that the boundaries of the area modelled should be located sufficiently distant from the excavation to minimize errors in the area of interest around the opening.

In the application of finite element methods in underground coal mining openings for example, this is normally achieved by placing the boundaries at a distance of approximately ten times the length of a longwall face, on account of the longitudinal geometry of the underground geometry. Furthermore, in the case of tunnelling operations, the boundary is usually located at a minimum distance of five tunnel radii from the walls of the opening, since it has been verified from closed-form solutions that at this distance the stress concentrations are insignificant (Obert and Duvall, 1967).

Kulhawy (1974) in particular, has compared closed-form solutions of a hole in a circular plate with finite element results, and found that when the mesh boundary is located at six tunnel radii, the stresses and displacements from the finite element solution are accurate to within 10%.

Since the perfect finite element solution can only be achieved by setting the boundaries truly at infinity, the discrepancies of 10% for setting the boundaries at a distance of six tunnel radii appears to be an acceptable approximation.

It is interesting to note that further larger increases in the distance to the boundary may only slightly decrease the inevitable inaccuracies.

10.5 Modelling of underground excavation support systems

In designing a tunnel support system, the engineer normally considers one or a combination of the following supports: concrete lining, shotcrete, steel sets, rockbolts and dowels (Brady and Brown, 1985).

Shotcrete is pneumatically applied concrete used to provide passive support to the surface of a ground excavation. Steel sets or steel arches are cold rolled from standard steel sections and are usually fabricated in two or more pieces to facilitate storage, handling and installation. Steel sets are used where high load-carrying capacity elements are required to support tunnels. Such sets may be fabricated from any steel section ranging from light sections (4 I 7.7) to heavy sections (14 WF 211) or larger. A tensioned rockbolt consists of an anchorage, a shank, a simple or deformable face plate and a tightening nut. Early rockbolt anchorages were of the mechanical slot-and-wedge or expansion shell type. Today, anchorages consisting of Portland cement grout or resin are generally used in ground support applications, since they have proved to be more reliable and permanent. Dowels are similar to rockbolts and made of same materials but are not pre-tensioned and are grouted along their full length.

For the specific application of steel arches or steel sets, engineers normally use as a standard reference the work by Proctor and White (1977) which provides the most detailed account available of the materials and techniques used for steel support.

However, for the case of concrete lining or shotcrete and rockbolt or dowel support, the finite element method is regularly used to model the interaction between ground and support systems.

Since the excavation of a tunnel is advancing parallel to its axis, care must be exercised when using a two-dimensional finite element analysis, to take into consideration the influence of the excavation process along the third dimension. It has been established (Baudendistel, 1985) that for tunnel excavations, in typical cases of elasto-plastic ground behaviour, the deformations at a cross-section in the ground ahead of the excavation begin to take place when the tunnel face is 0.5–1 tunnel diameter away. As soon as the advancing tunnel face will reach the cross-section under consideration, it is expected that 25–30% of the total deformation that may ever occur will take place. The remainder of the deformation is anticipated to occur by the time the tunnel face advances further to a distance of 1.5–2 tunnel diameters. In addition, account must be taken of the time delay between the ground excavation phase at a specific cross-section and the introduction of the support system. As a result, nearly 40–60% of the total tunnel deformation has already occurred by the time the support is installed.

In line with the above remarks, when modelling a support system, allowances must be made to permit the elasto-plastic ground to deform prior to the introduction of the support elements. This may be achieved by analysing the problem in consecutive load steps (using the INCREMENTAL module and the RESTART facility), whereby the first load step will allow the ground surrounding the tunnel to relax up to, for example, 60% of the deformation corresponding to the creation of the opening. The remaining 40% of the relaxation will be implemented once the elements corresponding to the support system have been introduced in the model using the CONTROL option MORE.DATA.

10.5.1   Concrete lining or shotcrete

The analysis of shotcrete support is identical to the one employed for *in situ* cast concrete, the only difference being the thickness of the lining (shotcrete is normally limited to a thickness ranging between 0.25 and 0.02 of the tunnel diameter for tunnels in the 5–10 m diameter category).

In a typical underground excavation sequence, shotcrete is installed early on in the process and since its strength increases with time, the shotcrete lining is expected to be stressed while still weak.

To model such a sequence, the ground may be allowed to relax, for example by 50% while the tunnel is unsupported, 25% while shotcrete is still weak (using say a Young's modulus of 5 GPa) and 25% assuming that the shotcrete has fully hardened (whereby the Young's modulus may be increased to 15 GPa).

10.5.2   Rockbolts and dowels

In certain finite element packages, specifically designed for geotechnical applications, individual tensioned rockbolts may be modelled using a spring element of zero dimension but to which a stiffness may be assigned. However, the SPRINGS module available in PAFEC-FE does not offer such a facility. Instead, SPRINGS are essentially employed whenever a discrete stiffness needs to be introduced at a joint foundation or in some complicated piece of structure which is not modelled in detail.

When PAFEC-FE is used, the pre-tensioning effect exercised by rockbolts may be modelled by using the DISPLACEMENTS.PRESCRIBED module for the nodes corresponding to the anchorage end and the face plate end of the rockbolt.

10.6   Example of a near surface excavation using PAFEC-FE

10.6.1   Description of the problem

In this example a 10 m deep by 30 m wide rectangular cross-section excavation, will be analysed to show how PAFEC-FE may be used to assess the stability of the excavation by calculating the stress and displacement distribution in the ground.

In order to use PAFEC-FE for this example several assumptions were made concerning the geometrical modelling of the excavation, including the stress–strain relationship of the geological materials involved, the initial geostatic stresses in the ground and the simulation of the excavation process.

The geometric idealization of the 10 m deep excavation is shown in Figure 10.1. The main ground formation is overlaid by a layer of gravel and a layer of ground surface drift material each 3 m thick. Since the excavated trench is taken to extend considerably along its longitudinal axis, a two-dimensional mesh idealization is used employing a plane strain analysis. Furthermore, because of the existing symmetry of the excavation with respect to the vertical plane passing through its centreline, it is sufficient to model only half of the structure (Figure 10.1).

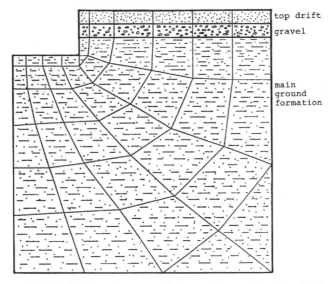

top drift

gravel

main
ground
formation

Figure 10.1 Geometric idealization of a near surface ground excavation, showing the ground formation types and the finite element mesh used (plane strain analysis).

With reference to the constitutive response of the ground formations account has to be taken that in a shallow excavation, the resulting stress concentrations are low and an isotropic linear elastic behaviour is to be expected. Even if this assumption does not correspond to the true stress–strain behaviour of the materials involved, it may be used to provide a first-order estimate of the range of the associated stresses and displacements and it may also form an adequate basis for extrapolation to other designs.

In this particular example the geostatic stresses will be introduced by making use of the gravity approach and the simulation of the excavation process will be implemented by employing the stress reversal technique, in which case two consecutive PAFEC-FE analyses will be necessary.

During the first analysis, the conditions before the excavation are simulated and the forces acting on the elements adjacent to the excavation boundary due to the geostatic stresses are determined using the module EXTERNAL.FORCES. In the second analysis the excavation is simulated by eliminating the appropriate elements, and the forces obtained from the first run are applied to the excavation boundary after their sign is reversed.

The PAFEC-FE data for the two analyses, enhanced with an extended set of interspaced comments are presented in the following section together with the associated graphical output.

10.6.2   PAFEC-FE input data

```
C   PAFEC-FE data for the 1st analysis. Unexcavated phase.
C
TITLE   NEAR SURFACE EXCAVATION IN OF A TRENCH (PHASE 1)
C
C   The coordinates of the nodes corresponding to the unexcavated structure
C   are given below expressed in metres.
C   Note that the origin of axes is node no. 54, located at a depth of 58 m
C   and lying on the plane of symmetry.
C
C   There is only need to specify the corner nodes of the elements, since
C   PAFEC-FE will calculate automatically the coordinates of the mid-side
C   nodes.
C
NODES
```

NODE.NUMBER	X	Y
1	0.00	48.00
2	3.00	48.00
3	6.67	48.00
4	11.00	48.00
5	15.00	48.00
6	15.00	52.00
7	15.00	55.00
8	15.00	58.00
9	0.00	45.67
10	3.00	45.67
11	6.67	45.67
12	11.00	45.67
13	15.00	45.67
14	17.00	48.00
15	17.33	52.00
16	17.33	55.00
17	17.33	58.00
18	0.00	40.67
19	3.33	40.67
20	8.33	40.67
21	13.33	40.67
22	18.00	42.00
23	21.67	46.33
24	22.67	52.00
25	22.67	55.00
26	22.67	58.00

27	0.00	32.67
28	5.00	33.00
29	11.33	33.33
30	19.33	33.67
31	26.00	37.33
32	30.67	44.67
33	31.00	52.00
34	31.00	55.00
35	31.00	58.00
36	0.00	23.33
37	8.00	23.33
38	16.00	24.33
39	26.00	26.00
40	35.00	32.00
41	40.00	43.00
42	40.00	52.00
43	40.00	55.00
44	40.00	58.00
45	0.00	12.67
46	11.33	13.33
47	23.67	14.00
48	35.67	17.00
49	47.00	28.00
50	49.00	42.67
51	49.00	52.00
52	49.00	55.00
53	49.00	58.00
54	0.00	0.00
55	16.00	0.00
56	33.00	0.00
57	45.67	9.33
58	58.00	24.00
59	58.00	42.67
60	58.00	52.00
61	58.00	55.00
62	58.00	58.00
63	58.00	0.00
64	0.00	58.00
65	3.00	58.00
66	6.67	58.00
67	11.00	58.00
68	0.00	55.00
69	3.00	55.00
70	6.67	55.00
71	11.00	55.00
72	0.00	52.00
73	3.00	52.00
74	6.67	52.00
75	11.00	52.00

C
C   There are three geological materials (drift, gravel and the main ground
C   formation) for which different PROPERTIES are assigned, namely 1, 2
C   and 3.
C
C   Two different types of isoparametric elements are used to accommodate
C   the geometry of the finite element mesh, namely the eight noded
C   curvilinear quadrilateral element (36210) and the six noded curvilinear
C   triangular element (36110).
C
C   To define the topology of the elements, only the corner nodes are used,
C   since no curved sides exist in the mesh.
C
ELEMENTS

NUMBER	ELEMENT.TYPE	PROPERTIES	TOPOLOGY			
1	36210	1	7	16	8	17
2	36210	1	16	25	17	26
3	36210	1	25	34	26	35
4	36210	1	34	43	35	44
5	36210	1	43	52	44	53
6	36210	1	52	61	53	62
7	36210	2	6	15	7	16
8	36210	2	15	24	16	25
9	36210	2	24	33	25	34
10	36210	2	33	42	34	43
11	36210	2	42	51	43	52
12	36210	2	51	60	52	61
13	36210	3	5	14	6	15
14	36210	3	14	23	15	24
15	36210	3	23	32	24	33
16	36210	3	32	41	33	42
17	36210	3	41	50	42	51
18	36210	3	50	59	51	60
19	36110	3		5	14	13
20	36210	3	13	22	14	23
21	36210	3	22	31	23	32
22	36210	3	31	40	32	41
23	36210	3	40	49	41	50
24	36210	3	49	58	50	59
25	36210	3	4	12	5	13
26	36210	3	12	21	13	22
27	36210	3	21	30	22	31
28	36210	3	30	39	31	40
29	36210	3	39	48	40	49
30	36210	3	48	57	49	58
31	36110	3		57	63	58
32	36210	3	3	11	4	12

33	36210	3	11	20	12	21
34	36210	3	20	29	21	30
35	36210	3	29	38	30	39
36	36210	3	38	47	39	48
37	36210	3	47	56	48	57
38	36110	3		56	63	57
39	36210	3	2	10	3	11
40	36210	3	10	19	11	20
41	36210	3	19	28	20	29
42	36210	3	28	37	29	38
43	36210	3	37	46	38	47
44	36210	3	46	55	47	56
45	36210	3	1	9	2	10
46	36210	3	9	18	10	19
47	36210	3	18	27	19	28
48	36210	3	27	36	28	37
49	36210	3	36	45	37	46
50	36210	3	45	54	46	55
51	36210	1	68	69	64	65
52	36210	1	69	70	65	66
53	36210	1	70	71	66	67
54	36210	1	71	7	67	8
55	36210	2	72	73	68	69
56	36210	2	73	74	69	70
57	36210	2	74	75	70	71
58	36210	2	75	6	71	7
59	36210	3	1	2	72	73
60	36210	3	2	3	73	74
61	36210	3	3	4	74	75
62	36210	3	4	5	75	6

```
C
C    The PLATES.AND.SHELLS module is used to link the PROPERTIES
C    specified in the ELEMENTS module with the MATERIAL.NUMBER
C    used in the MATERIAL module.
C
PLATES.AND.SHELLS
PLATE.NUMBER        MATERIAL.NUMBER
    1               11
    2               12
    3               13
C
C    In the MATERIAL module the Young's modulus (E), the Poisson's ratio
C    (NU) and the density (RO), are specified for the top drift
C    (MATERIAL.NUMBER 11), the gravel (MATERIAL.NUMBER 12) and
C    the main ground formation (MATERIAL.NUMBER 13).
C
C    The unit used for Young's modulus is Pa and the density is expressed in
C    kg/m³
```

```
C
MATERIAL
MATERIAL.NUMBER          E      NU      RO
    11                 4E+07   0.350   1800
    12                 4E+07   0.350   1900
    13                 8E+07   0.489   2000
C
C   In the GRAVITY module the sign in the YGVALUE is negative,
C   indicating that gravity acts in the vertical downwards direction.
C
GRAVITY
XGVALUE        YGVALUE
    0            −1
C
C   To ensure that the nodes lying on the plane of symmetry are restricted
C   from moving in the horizontal direction, all nodes lying on the plane that
C   passes through NODE.NUMBER 54 and is normal to the x-axis (PLANE
C   1), are prevented from moving along the x-direction (DIRECTION 1).
C   Similarly, the right side boundary is also prevented from moving along the
C   horizontal direction, by employing this time node 63 instead of node 54.
C   In this way the horizontal geostatic stresses are allowed to develop as a
C   function of the Poisson's ratio of the ground.
C
C   Furthermore there is a need to restrict the lower boundary of the mesh
C   from moving in the vertical direction, so a restraint is introduced to
C   prevent any node lying on the plane that passes from NODE.NUMBER 54
C   and is normal to y-axis (PLANE 2), to move along the y-direction
C   (DIRECTION 2).
C
RESTRAINTS
NODE.NUMBER        PLANE   DIRECTION
    54               1         1
    54               2         2
    63               1         1
C
C   Two drawings are requested before the displacements and stresses
C   calculations: a blown up mesh drawing with solid lines for the boundaries
C   of the elements (TYPE.NUMBER 1), where the element numbers are
C   shown (INFORMATION.NUMBER 3) with small arrows drawn indicating
C   the applied restraints (INFORMATION.NUMBER 5). Note that the
C   INFORMATION.NUMBER is 35, i.e. a compound number of 3 and 5.
C
C   The second drawing is the mesh of the structure drawn with solid line
C   throughout (TYPE.NUMBER 2) with the node numbers printed
C   (INFORMATION.NUMBER 2).
C
```

```
IN.DRAW
TYPE.NUMBER  INFORMATION.NUMBER
     1              35
     2               2
C
C  Following the displacements and stresses calculations, two drawings are
C  requested: the displaced shape (PLOT.TYPE 2) and the in-plane stress
C  vectors (PLOT.TYPE 20).
C
OUT.DRAW
PLOT.TYPE
     1
    20
C
C  The EXTERNAL.FORCES module calculates and prints the forces acting
C  on the elements whose numbers are given in the LIST.
C
EXTERNAL.FORCES
LIST
    45  39  32  25  19  13  7  1
C
C  The CONTROL facility is used to specify plane strain analysis and to
C  indicate that stress averaging across different material types is to be
C  performed.
C
CONTROL
PLANE.STRAIN
C
C  Since PAFEC-FE does not perform stress averaging across different
C  material types, the supplied source must be modified. Inspection of the
C  stressing routines indicated that a small number of changes to PAFEC-FE
C  subroutine R70632 was necessary. A section of the subroutine with the
C  modification is shown in Figure 10.2. The modified source is incorporated
C  into the PAFEC-FE system for this analysis using the USE. option. In this
C  example it is assumed that the whole of the modified subroutine has been
C  copied to a disk file named 'average'.
C
C
USE.average
C
CONTROL.END
C
END.OF.DATA
```

Comparison of the next set of PAFEC-FE data, used in the second analysis, with the data of the first analysis, highlights the following important differences:

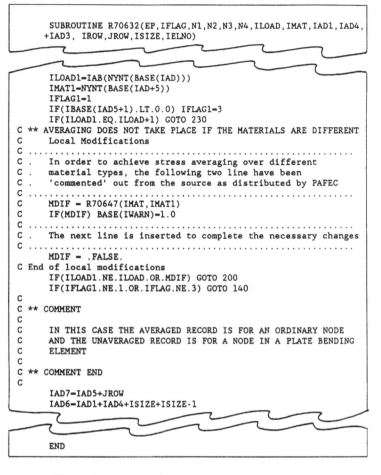

```
        SUBROUTINE R70632(EP,IFLAG,N1,N2,N3,N4,ILOAD,IMAT,IAD1,IAD4,
        +IAD3, IROW,JROW,ISIZE,IELNO)



        ILOAD1=IAB(NYNT(BASE(IAD)))
        IMAT1=NYNT(BASE(IAD+5))
        IFLAG1=1
        IF(IBASE(IAD5+1).LT.0.0) IFLAG1=3
        IF(ILOAD1.EQ.ILOAD+1) GOTO 230
C ** AVERAGING DOES NOT TAKE PLACE IF THE MATERIALS ARE DIFFERENT
C     Local Modifications
C ...................................................................
C .   In order to achieve stress averaging over different
C .   material types, the following two line have been
C .   'commented' out from the source as distributed by PAFEC
C ...................................................................
C     MDIF = R70647(IMAT,IMAT1)
C     IF(MDIF) BASE(IWARN)=1.0
C ...................................................................
C .   The next line is inserted to complete the necessary changes
C ...................................................................
        MDIF = .FALSE.
C End of local modifications
        IF(ILOAD1.NE.ILOAD.OR.MDIF) GOTO 200
        IF(IFLAG1.NE.1.OR.IFLAG.NE.3) GOTO 140
C
C ** COMMENT
C
C     IN THIS CASE THE AVERAGED RECORD IS FOR AN ORDINARY NODE
C     AND THE UNAVERAGED RECORD IS FOR A NODE IN A PLATE BENDING
C     ELEMENT
C
C ** COMMENT END
C
        IAD7=IAD5+JROW
        IAD6=IAD1+IAD4+ISIZE+ISIZE-1



        END
```

Figure 10.2   Section of PAFEC-FE subroutine R70632.

- Since in this analysis the excavation is formed by removing elements, less elements are specified (50 instead of 62) and the number of nodes used in the mesh is smaller (63 instead of 75).
- The LOADS module is used to apply the reversed sign forces calculated by the EXTERNAL.FORCES module in the first analysis.

```
C   PAFEC-FE data for the 2nd analysis. Excavation phase.
C
TITLE   NEAR SURFACE EXCAVATION OF A TRENCH (PHASE 2)
C
NODES
NODE.NUMBER         X       Y
     1             0.00    48.00
     2             3.00    48.00
```

3	6.67	48.00
4	11.00	48.00
5	15.00	48.00
6	15.00	52.00
7	15.00	55.00
8	15.00	58.00
9	0.00	45.67
10	3.00	45.67
11	6.67	45.67
12	11.00	45.67
13	15.00	45.67
14	17.00	48.00
15	17.33	52.00
16	17.33	55.00
17	17.33	58.00
18	0.00	40.67
19	3.33	40.67
20	8.33	40.67
21	13.33	40.67
22	18.00	42.00
23	21.67	46.33
24	22.67	52.00
25	22.67	55.00
26	22.67	58.00
27	0.00	32.67
28	5.00	33.00
29	11.33	33.33
30	19.33	33.67
31	26.00	37.33
32	30.67	44.67
33	31.00	52.00
34	31.00	55.00
35	31.00	58.00
36	0.00	23.33
37	8.00	23.33
38	16.00	24.33
39	26.00	26.00
40	35.00	32.00
41	40.00	43.00
42	40.00	52.00
43	40.00	55.00
44	40.00	58.00
45	0.00	12.67
46	11.33	13.33
47	23.67	14.00
48	35.67	17.00
49	47.00	28.00
50	49.00	42.67

51	49.00	52.00
52	49.00	55.00
53	49.00	58.00
54	0.00	0.00
55	16.00	0.00
56	33.00	0.00
57	45.67	9.33
58	58.00	24.00
59	58.00	42.67
60	58.00	52.00
61	58.00	55.00
62	58.00	58.00
63	58.00	0.00

C
ELEMENTS

NUMBER	ELEMENT.TYPE	PROPERTIES	TOPOLOGY			
1	36210	1	7	16	8	17
2	36210	1	16	25	17	26
3	36210	1	25	34	26	35
4	36210	1	34	43	35	44
5	36210	1	43	52	44	53
6	36210	1	52	61	53	62
7	36210	2	6	15	7	16
8	36210	2	15	24	16	25
9	36210	2	24	33	25	34
10	36210	2	33	42	34	43
11	36210	2	42	51	43	52
12	36210	2	51	60	52	61
13	36210	3	5	14	6	15
14	36210	3	14	23	15	24
15	36210	3	23	32	24	33
16	36210	3	32	41	33	42
17	36210	3	41	50	42	51
18	36210	3	50	59	51	60
19	36110	3		13	14	5
20	36210	3	13	22	14	23
21	36210	3	22	31	23	32
22	36210	3	31	40	32	41
23	36210	3	40	49	41	50
24	36210	3	49	58	50	59
25	36210	3	4	12	5	13
26	36210	3	12	21	13	22
27	36210	3	21	30	22	31
28	36210	3	30	39	31	40
29	36210	3	39	48	40	49
30	36210	3	48	57	49	58
31	36110	3		57	63	58
32	36210	3	3	11	4	12

33	36210	3	11	20	12	21
34	36210	3	20	29	21	30
35	36210	3	29	38	30	39
36	36210	3	38	47	39	48
37	36210	3	47	56	48	57
38	36110	3		56	63	57
39	36210	3	2	10	3	11
40	36210	3	10	19	11	20
41	36210	3	19	28	20	29
42	36210	3	28	37	29	38
43	36210	3	37	46	38	47
44	36210	3	46	55	47	56
45	36210	3	1	9	2	10
46	36210	3	9	18	10	19
47	36210	3	18	27	19	28
48	36210	3	27	36	28	37
49	36210	3	36	45	37	46
50	36210	3	45	54	46	55

C
PLATES.AND.SHELLS

PLATE.NUMBER	MATERIAL.NUMBER
1	11
2	12
3	13

C
MATERIAL

MATERIAL.NUMBER	E	NU	RO
11	4E+07	0.350	1800
12	4E+07	0.350	1900
13	8E+07	0.489	2000

C
C Here the loads (expressed in N) that have been calculated in the 1st
C analysis, are applied on the boundary of the excavation after their sign is
C reversed.
C
LOADS

NODE.NUMBER	DIRECTION.OF.LOAD	VALUE.OF.LOAD
1	1	−69630.187
1	2	82257.562
64	1	−13.402
64	2	420000.0
2	1	−15.5
2	2	183280.562
65	1	33.788
65	2	514000.0
3	1	−4.063
3	2	220000.000

66	1	−24.783
66	2	607000.0
4	1	−40.625
4	2	229000.0
67	1	52.509
67	2	561000.0
5	1	−119984.0
5	2	95543.695
68	1	−378000.0
68	2	56565.32
6	1	−98821.121
6	2	−24612.649
69	1	−87154.937
69	2	43422.925
7	1	−28525.835
7	2	−21138.199
70	1	−28524.503
70	2	41143.285
8	1	−0.267
8	2	−10285.828

C
RESTRAINTS

NODE.NUMBER	PLANE	DIRECTION
54	1	1
54	2	2
63	1	1

C
IN.DRAW

TYPE.NUMBER	INFORMATION.NUMBER
1	35
2	2

C
C This time some additional drawings are requested, the maximum (most
C tensile) principal in-plane stress contours (PLOT.TYPE 30), the minimum
C (most compressive) principal in-plane stress contours (PLOT.TYPE 31)
C and the von Mises stress contours (PLOT.TYPE 35)
C
OUT.DRAW

PLOT.TYPE
1
20
30
31
35

C
CONTROL
PLANE.STRAIN
C

```
C   The REDUCED.OUTPUT option is used in the second analysis in order
C   to lessen the volume of output. Having successfully completed the first
C   analysis there is no need to get once more such items as the lists of nodal
C   coordinates and element nodes.
C
REDUCED.OUTPUT
C
C   Once more the modified form of R70632 is used to achieve stress
C   averaging over different material types.
C
USE.average
C
CONTROL.END
C
END.OF.DATA
```

10.6.3 Discussion of results

The interpretation of the PAFEC-FE results revolves around the plots that
PAFEC-FE produces by means of the IN.DRAW and OUT.DRAW modules.
From these it will become apparent that the numerical results are indeed
acceptable.

In Figure 10.3 part of the finite element idealization of the unexcavated
ground is shown by blowing up the mesh and drawing each element with
solid lines. Since the number of every element is given in the graphics
output, it is clear that this is very useful in allowing the user to inspect the
correctness of the element topology. It is in this area that most of the input
data errors occur. In the same drawing the user can verify the requested
restraints by checking the presence and direction of the small arrows on the
boundaries of the structure.

Figure 10.4 shows part of the finite element mesh where the node numbers
are given for every element and which is mainly used as a reference drawing
to locate the relative position of nodes.

The displaced shape of the ground structure under the action of the geostatic stresses is shown in Figure 10.5, where the original shape is drawn
with solid lines and the deformed shape is shown in broken lines. Some
implementations of PAFEC-FE, permit the deformed shape to be drawn in
coloured broken lines, in which case it is very easy to distinguish the new
geometry of the structure. What is important at this stage is to use the information given in the displaced shape to verify that the geostatic loading
has been applied in a uniform manner by checking that the deformed shape
has no irregularities. Indeed, it can be seen clearly in Figure 10.5 that the
ground surface retains its horizontal disposition along the entire length of
the cross-section, indicating a satisfactory application of the GRAVITY
module.

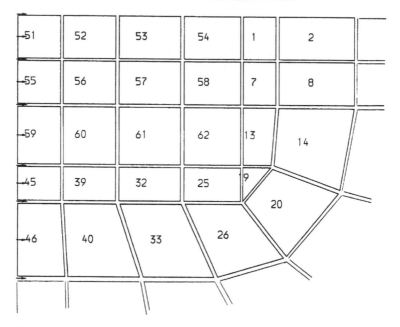

Figure 10.3 Part of the finite element mesh of the near surface ground excavation prior to excavation phase. The mesh is blown up, the element numbers are shown and the applied restraints at the boundaries are indicated by small arrows.

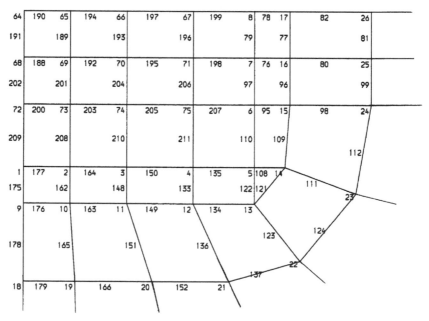

Figure 10.4 Part of the finite element mesh of the near surface ground excavation prior to excavation phase (node numbers are shown).

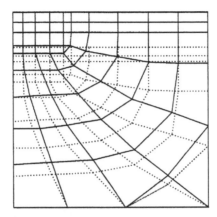

Figure 10.5 Deformed shape of the ground prior to excavation phase, under the influence of geostatic stresses (plane strain analysis).

In addition to the deformed shape it is useful to get an idea of the distribution of the principal stresses in the structure resulting from the application of the geostatic stress field. For this, we may use Figure 10.6, where the in-plane stress vectors are drawn at each node. The advantage of the stress vector drawing is the clear illustration of the magnitude, nature and, most important, orientation of the two principal stresses. In this particular case, as was expected, Figure 10.6 shows the linear increase of the geostatic stresses with depth and the ideal orientation of the vertical and horizontal stress components.

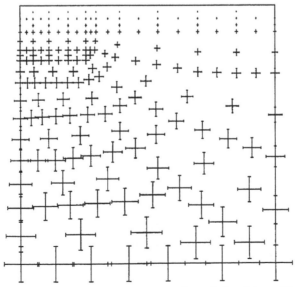

Figure 10.6 In-plane stress vectors for plane strain analysis of the solid ground prior to excavation.

In the second analysis where the excavation has been created, Figures 10.7 and 10.8 repeat the information given by Figures 10.3 and 10.4 by demonstrating all the relevant geometric information with reference to the position of the nodes and the topology of the elements.

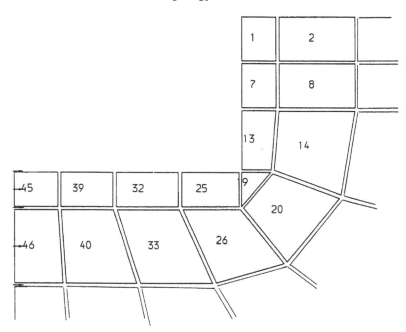

Figure 10.7 Part of the finite element mesh of a near surface ground excavation following excavation phase. The mesh is blown up, the element numbers are shown and the applied restraints at the boundaries are indicated by small arrows.

The stress vectors shown in Figure 10.9 show the marked influence that the excavation has on the stress distribution in the ground. Figure 10.9 shows the high stress concentrations developing in the bottom corner of the excavation and the characteristic disturbance in the orientation of the principal stress directions from vertical and horizontal (compare with Figure 10.6). to a pattern corresponding to typical stress trajectories around ground excavations (see Appendix 3 in Hoek and Brown, 1980).

The other three drawings, Figures 10.10, 10.11 and 10.12 give stress distribution information by showing stress contour diagrams around the excavation. Here it can be seen that the magnitudes of both the major and the minor principal stresses are affected mainly near the walls of the excavation while further into the ground this effect diminishes sharply. This is not unexpected since the modelled trench is a near surface excavation and the stresses involved are relatively low in magnitude.

An interesting conclusion derived from Figures 10.10 and 10.11 is the absence of tensile stresses which indicates good stability conditions since

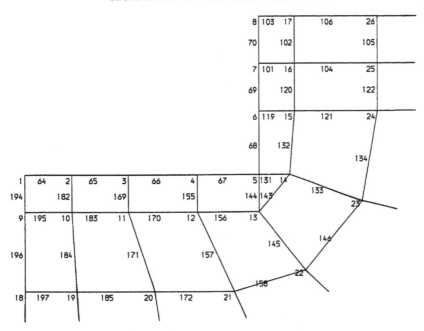

Figure 10.8 Part of the finite element mesh of the ground following the excavation phase of a near surface excavation. (node numbers are shown).

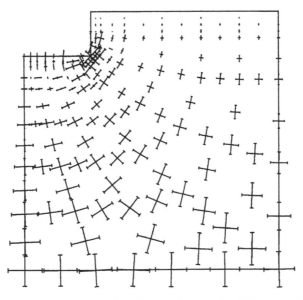

Figure 10.9 In-plane stress vectors for plane strain analysis of the ground following the excavation phase.

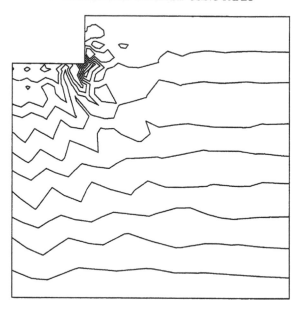

Figure 10.10 Maximum (most positive) principal in-plane stress contours for plane strain analysis of a near surface ground excavation.

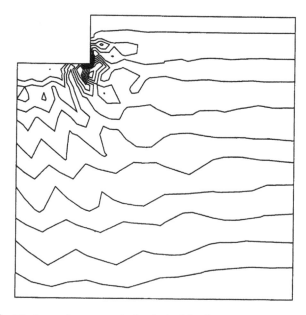

Figure 10.11 Maximum (most negative) principal in-plane stress contours for plane strain analysis of a near surface ground excavation.

any tension is expected to result in failure of the geological materials involved. The contours given in Figure 10.12 correspond to the distribution of the von Mises stress component equivalent to the octahedral shear stress normally used as a basis for the investigation of possible failure conditions under compressive stresses.

Figure 10.12 Stress contours of the von Mises stress component in the ground surrounding the excavation.

References

M. Baudendistel, Significance of the unsupported span in tunnelling, *Proc. Tunnelling 1985* (1985) 103–108.

B.H.G. Brady and E.T. Brown, *Rock Mechanics for Underground Mining*, Allen & Unwin, London (1985) 527pp.

I.M. Breckles, A study of the distribution and magnitude of strata loading around mine roadways and tunnels, Ph.D. Thesis, Department of Mining Engineering, University of Nottingham (1978).

I.M. Breckles, C. McCaul and B.N. Whittaker, Comparison between measured and simulated displacements that result from tunnel and shaft intersection in lower chalk, *IMM Trans.* **90** (1981) A34–A46.

A. Burt, ed., *Computer Methods in Tunnel Design*, The Institution of Civil Engineers, London (1978) 196pp.

R.E. Goodman, *Methods of Geological Engineering in Discontinuous Rock*, West, St. Paul (1976).

E. Hoek and E.T. Brown, *Underground Excavations in Rock*, Institution of Mining and Metallurgy, London (1980) 527pp.

F.H. Kulhawy, Finite element modelling criteria for underground openings in rock, *Int. J. Rock Mech. Min. Sci.* **11** (1974) 465–472.

D.J. Naylor, G.N. Pande, B. Simpson and R. Tabb, *Finite Elements in Geotechnical Engineering*, Pineridge Press, Swansea (1981) 245pp.

L. Obert and W.I. Duvall, *Rock Mechanics and the Design of Structures in Rock*, Wiley, New York (1967) 650pp.

R.J. Proctor and T.L. White, *Rock Tunnelling with Steel Supports*, rev. edn., Commercial Shearing and Stamping Co., Youngstown (1977).

J.M. Reed, Mine roadway modelling methods, Ph.D. Thesis, Department of Mining Engineering, University of Leeds (1988).

O.C. Zienkiewicz, *The Finite Element Method*, 3rd edn., McGraw-Hill, London (1977).

11 Fracture mechanics analysis

11.1 Introduction

In engineering applications there is often a need to improve the quantitative understanding we may have about situations where crack propagation is a major concern. In such cases, the engineering discipline of fracture mechanics is the ideal analytical tool to offer a description of the transformation of an intact structural component into one broken by crack growth.

Strength failures of load bearing structures or components can be either of the yielding-dominant or fracture-dominant types. When yielding occurs and plastic behaviour prevails, the solution of structural problems is customarily dealt by employing the theory of elasto-plasticity (Nadai, 1950). On the other hand, fracture mechanics, which is the subject of this chapter, is concerned almost entirely with fracture-dominant failure.

In its most elementary form, fracture mechanics analysis relates the maximum permissible stress applied to a structure to the size and location of a crack contained within the structure. It may also predict the rate at which cracks grow to a critical size by environmental influences or by varying loads, this latter condition being known as fatigue. Furthermore, it may also be used successfully to determine the conditions for rapid propagation and arrest of moving cracks (Kanninen and Popelar, 1985).

Fracture mechanics is primarily used in engineering applications to predict and to prevent catastrophic failures occurring in components or structures made of materials such as metals, ceramics and plastics. Furthermore, fracture mechanics is becoming important in its direct application to the cracking of concrete (Wittman, 1983; Shah, 1985) and natural materials such as rocks (Karfakis et al., 1986).

In developing analytical solutions for fracture mechanics problems, use is normally made of the fundamental principles of continuum mechanics such as linear elasticity and plasticity. As a result, it is often appropriate to employ the term linear elastic fracture mechanics to indicate that linear elasticity laws have been used to derive the analytical approach adopted.

Linear elastic fracture mechanics as such has gained widespread use in engineering applications, since several investigators have successfully applied it to analyse the problem of crack propagation in materials. A full treatment of the subject of fracture mechanics is available in relevant texts (for example Ewalds and Wanhill, 1985; Knott, 1973). However, in order to provide a basis for understanding the concepts and expressions used in this chapter, a brief outline of the basic terms and concepts is presented.

11.2   Fracture toughness

Fracture mechanics may be considered as a development of the strength of materials approach, whereby the stress in a component or a structure is compared with some material strength value in order to ascertain whether failure will occur or not. In the vicinity of the tip of a crack, there are localized stresses which are much higher than the nominal stresses in that region. The degree of such stress concentration is quantified by the stress intensity factor, which in fracture mechanics is considered as the basic mathematical quantity utilized in assessing the ability of a material to counter fracturing, that is, to resist the initiation of cracks.

The stress intensity factor K is essentially a measure of the singularity of the stress field at a loaded crack tip and is closely related to the available energy release rate (Irwin, 1957). It is usually determined by analysis and its dimensions are stress × (crack length)$^{1/2}$, i.e. Pa m$^{1/2}$.

First, from linear elastic theory Irwin showed that the stresses in the vicinity of a crack tip take the form

$$\sigma_{ij} = \frac{K}{(2\pi r)^{1/2}} f_{ij}(\theta) + \ldots$$

where the subscripts i and j vary from 1 to 3 and where r and θ, are the cylindrical polar coordinates of a point with respect to the crack tip.

Since the stress intensity factor K describes the magnitude of the elastic crack tip stress field, dimensional analysis shows that K must be linearly related to stress and directly related to the square root of a characteristic length. It transpires that this characteristic length is the crack length (2α), in which case the general form of the stress intensity factor is given by

$$K = \sigma\sqrt{\pi\alpha} \cdot f(\frac{\alpha}{W})$$

where $f(\alpha/W)$ is a dimensionless parameter depending on the geometries of the specimen and the crack.

K can correlate the crack growth and fracture behaviour of materials provided that the crack tip stress field remains predominantly elastic. This correlating ability makes the stress intensity factor an extremely important fracture mechanics parameter.

All stress systems in the vicinity of a crack tip may be derived from three states of loading corresponding to the three possible modes of crack extension: mode I (opening), mode II (sliding) and mode III (tearing) as shown

schematically in Figure 11.1. The corresponding intensity factors are customarily denoted by employing the subscript I, II or III to indicate opening,
sliding or tearing mode, respectively.

Examination of the available literature on fracture mechanics indicates
that since mode I is the predominant stress situation in many practical cases,
more attention has been paid to the investigation of stress intensity factor
K_I in tension. However, for certain engineering problems the sliding mode
stress intensity factor K_{II} and the mixed mode stress intensity factor K_{III} are
equally dominant, and their consideration may be of particular importance.

By definition, the stress intensity factor K_I, is related only to the loading
and geometry of the structural element. It has also been shown (Irwin, 1957)
that the change in potential energy during infinitesimal crack extension is
related to stress intensity. Specifically,

$$-\frac{\partial U}{\partial \alpha} = G_I = \frac{K_I^2 (1 - \nu^2)}{E} \quad \text{plane strain problems}$$

$$= \frac{K_I^2}{E} \quad \text{plane stress problems}$$

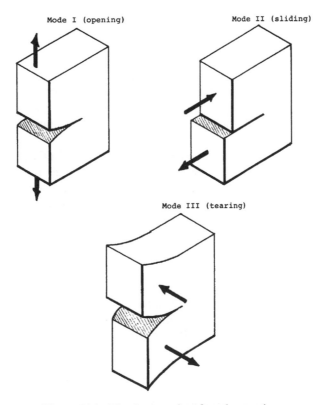

Mode I (opening)　　　　　Mode II (sliding)

Mode III (tearing)

Figure 11.1 The three modes of crack extension.

where K_I = opening mode stress intensity factor, α = the new crack surface area per unit width produced by infinitesimal crack extension, U = total potential energy of structural element, G_I = strain energy release rate, ν = Poisson's ratio and E = Young's modulus.

Equivalent expressions for K_{II} and K_{III} are available linking these intensity factors to the change in potential energy.

If the applied stress to a structure or a component containing a crack reaches a critical level, the crack is expected to propagate catastrophically. This occurs when K_I becomes equal to a critical value K_{Ic}, which is customarily taken as a material property, called fracture toughness. When K_I reaches K_{Ic}, catastrophic crack growth is assumed to occur. Thus, a structure or a component can be designed to be safe if K_I is kept below K_{Ic} and failure or fragmentation may be achieved if K_{Ic} is exceeded.

The fracture toughness of a material is a measure of the resistance to crack extension that may eventually result in structural failure and, like Young's modulus, it is a material constant. The fracture toughness also indicates the fracture energy consumption rate required to create new surfaces and is quantitatively expressed in units of Pa m$^{1/2}$ as the critical value that the stress intensity factor may assume.

11.3 *J* integral

The theory of linear elastic fracture mechanics, as already indicated, was originally developed to describe crack growth and fracture under essentially elastic conditions. To examine fracture behaviour beyond the elastic regime, use of the elastic-plastic fracture mechanics is made, although such methods are also limited.

The most widely used techniques to predict crack initiation, in elastic-plastic fracture mechanics, are either the crack opening displacement approach or the *J* integral approach of which the latter appears to be the most popular method. The *J* integral approach characterizes the crack in the presence of yielding in a way similar to that in which the stress intensity factor *K*, characterizes both the elastic fields and energy available for propagation of a crack in an elastic material.

Through the *J* integral, the concepts of linear elastic fracture mechanics may be extended into the post-yield regime. The *J* integral concept is based on an energy balance approach, and is a path independent integral expression representing a non-linear elastic energy release rate. The *J* integral expression was first introduced by Rice (1968) as

$$J = -\frac{\partial U_p}{\partial \alpha}$$

where U_p is the potential energy encompassing all the energy terms that may contribute to non-linear elastic behaviour. The quantity $\partial U_p/\partial \alpha$ represents change in stored energy, with the minus sign implying a decrease in the stored energy necessary to the release of crack driving energy J.

The path independency of the J integral expression allows calculation along a contour remote from the crack tip. Such a contour can be chosen to contain only elastic loads and displacements. Thus an elastic-plastic energy release rate can be obtained from an elastic calculation along a contour for which loads and displacements are known.

By definition the J integral is equal to $-(\partial U/\partial \alpha)$ for the linear elastic case. Thus by analogy the dimensions of J are (energy)/(length) per unit thickness of material, i.e. J/m^2 or N/m.

11.4 Application of finite element analysis in fracture mechanics

The principles of linear elastic fracture mechanics are well established and the concepts find increasing application in the design of engineering components against brittle fracture. An essential step in applying these concepts to the calculation of critical load levels or critical defect sizes is the accurate determination of stress intensity factors at complex structural configurations. Since exact solutions are limited to simple geometries and idealized boundary conditions, numerical methods such as finite element analysis must inevitably be utilized for a description of real situations.

Furthermore, in attempting to use fracture mechanics it will often be found that there is no standard stress intensity factor solution for the particular crack shape and structural component geometry under consideration. Consequently, approximate methods such as finite element calculations are used to obtain the required stress intensity factor.

It is the basic assumption of linear elastic fracture mechanics that the small displacement theory of linear elasticity provides a valid model even though infinite strains occur at a crack tip. In the application of linear elastic fracture mechanics to engineering problems, it is a fundamental objective to characterize the singular stress field at the crack tip in terms of the three stress intensity factors K_I, K_{II} and K_{III}. As each of these is associated with a $r^{-1/2}$ stress singularity, the strain energy in the region of the crack tip is finite so that the finite element energy method is immediately applicable and the usual convergence properties hold good.

As was indicated in section 6.15, in crack tip situations there is a need for strain singularity to occur at one end of the element around the tip of the crack. It may be shown (PAFEC-FE Theory manual) that the singularity in the strain expression for the element is of the order $r^{-1/2}$, where r is the distance from the crack tip. A desirable feature in the determination of stress intensity factor is that the $r^{-1/2}$ crack tip singularity can be modelled without the use of special elements around the crack tip.

The elements at the crack tip are not in any way special and the standard isoparametric elements may be used throughout the mesh. In fact, PAFEC-FE imposes proper stress singularity simply by enforcing the use of triangular elements around the crack tip and by moving the midside nodes of two adjacent sides from their usual position at the centre of each side to the quarter position.

Although the elements at the crack tip have reasonably accurate stiffness, the local values of stress and displacement in the elements adjacent to the crack tip are generally poor and can justifiably be ignored when calculating K_I. For this reason, elements attached to a crack tip node will not normally be stressed even if they appear in the STRESS.ELEMENTS module.

If, however, stresses are required at the crack tip node, the CONTROL option CRACK. STRESS should be included in the CONTROL module.

Stress intensity factors and J integral values are calculated after stress output in Phase 9. In order to obtain results for two-dimensional isoparametric elements, all elements containing crack tip nodes must be the triangular type 36110. For three-dimensional elements, all elements containing crack tip nodes must be the wedge type 37210 with the crack running along edges of the rectangular faces.

To obtain correct J integral values, it is important to ensure that stresses are produced for elements at a 'two element' radius from the crack tip. If elements attached to a crack tip node are stressed, average nodal values at the interface between the 'one element' and 'two element' radius will be used.

11.5   PAFEC-FE example

The present example is concerned with the examination of the K_I feature of thick rings containing a single internal edge crack. Such an approach is usually undertaken when aiming to design a specimen for which K_I remains constant over a large range of non-dimensional crack lengths $[\alpha/(R_o - R_i)]$ where R_o and R_i are the outer and inner radii of the ring (see Figure 11.2).

Such specimens are useful in parametric crack-growth studies, corresponding for instance to experiments aimed at studying the effect of an aggressive environment upon crack growth where the influence of changing K_I needs to be excluded.

For the cylindrical geometries considered here, symmetry requires only one-half of the test-piece (ring) to be analysed for one notch, the crack being modelled by a series of unrestrained nodes along the axis of symmetry. This ring specimen with a single notch is considered under plane stress conditions loaded by a line load above the crack plane.

The finite element mesh used in the analysis is shown in Figure 11.3. A total of 430 nodes and 118 elements were used and since mild steel was

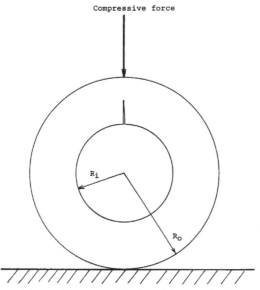

Figure 11.2   Ring specimen with outer radius R_o and inner radius R_i containing a radial crack and subjected to compressive diametric loading.

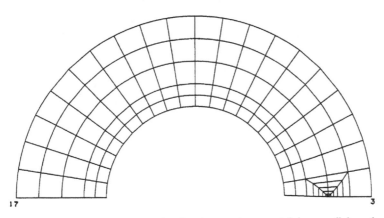

Figure 11.3   Finite element mesh of a ring specimen containing a radial crack.

used as material by default, Poisson's ratio was specified by the standard materials of PAFEC-FE as 0.3.

The PAFEC-FE data is presented in section 11.6. The TOLERANCE is reset from the (default) 10E–4 to 10E–6 in order to prevent PAFEC-FE forcing nodes which are very close to coincide.

Because of the cylindrical symmetry of the problem, AXIS is set to 3 in the NODES module, to facilitate the specification of the coordinates. The x-coordinates are expressed in metres and the y-coordinates in degrees.

PAFBLOCKS are used to specify the mesh; the first four blocks are TYPE = 1 and use quadrilateral 36210 elements. In effect, these four blocks define fully the geometry of the half ring but there is a need for greater concentration of elements near the tip of the crack. For this reason, two more blocks of TYPE = 5 (block numbers 5 and 6) are defined, so that they can be used with the REFERENCE.IN.PAFBLOCK facility to improve the density of the elements. Furthermore, at the actual tip of the crack, PAFEC-FE is expected to have triangular elements and consequently two more PAFBLOCKS are specified (block numbers 7 and 8) which employ triangular 36110 elements.

The effect of the MESH specifications as given in the data list, can be seen in Figure 11.3.

A compressive line load of 2 N is imposed above the crack plane (at node 3 shown in Figure 11.3) by using the LOADS module. The VALUE.OF.LOAD is negative, while the DIRECTION is specified as 1 since the load acts parallel to the x-axis and towards the origin of coordinates.

In the RESTRAINTS module the conditions of symmetry are satisfied by making sure that all the nodes lying on the line that passes through node 17 (see Figure 11.3) and parallel to the x-axis cannot move along the y-axis direction. Support is given to the lower end of the ring specimen by preventing node 17, which lies in the periphery of the ring and anti-diametrically to the crack, from moving along the x-axis direction.

The only graphical output available from a PAFEC-FE fracture analysis is from the IN.DRAW plots, and this serves to check the correctness of the mesh. Two plots have been requested: one not blown up with solid lines throughout (TYPE.NUMBER = 2) and nothing extra (INFORMA-TION.NUMBER = 0) as shown in Figure 11.3 and one with blown up, solid boundaries of elements (TYPE.NUMBER = 1) where the element numbers are also plotted (INFORMATION.NUMBER = 3) shown partially in Figure 11.4. To attempt to plot all the node numbers in the vicinity of the crack,

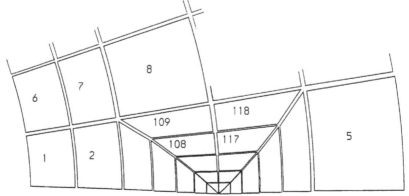

Figure 11.4 Part of the finite element idealization of a ring specimen containing a radial crack. Elements are shown numbered and blown up.

although a very important output for checking purposes, is often impractical because of the proximity of the nodes. Such plots are liable to be very cluttered even if the hardware available has enlarging facilities. An alternative strategy is to use the SELECT.DRAW module, which allows specific regions to be plotted in conjunction with the IN.DRAW module.

The results of the fracture analysis are given at the end of Phase 9 under the heading FRACTURE CALCULATIONS reproduced below.

***** FRACTURE CALCULATIONS *****

STRESS INTENSITY FACTORS	J-INTERGAL				AND J-INTEGRAL
		INTENSITY FACTORS			GIVEN BY STRESS
CRACK LOAD					J-INTEGNAL
TIP CASE	K-I	K-II	K-III	J-INTEGRAL	FROM
NODE					K-I, K-II, K-III
1 1	24.5	5.88	0.0	132.	0.303E–08

The particular problem of fracture analysis of the ring specimen was solved for different sizes of rings to ascertain whether there is any size effect influencing the fracture toughness measurements. For this purpose ring specimens were analysed with external diameter of 100 mm and 75 mm, and internal diameter of 50 mm and 45.5 mm resulting in a ratio (length of crack/width of ring) corresponding to 0.1, 0.25, 0.33 and 0.5.

The results of the analysis are customarily represented as non-dimensional or normalized stress intensity factors:

$$Y = \frac{K_I\, B\, \sqrt{R_o}}{P}$$ (Jones, 1974; Ahmad and Ashbaugh, 1982)

or

$$Y = \frac{K_I\, B\, (R_o - R_i)}{P\, (\pi\, \alpha)^{1/2}}$$ (Thompson *et al.*, 1984)

where B = thickness of ring specimen, R_i and R_o = the inner and outer radii of the ring and P = applied compressive load.

For the present application, the first non-dimensional convention was adopted and the results of the various analyses are given in Table 11.1.

Table 11.1

R_i/R_o	$\alpha/(R_o - R_i)$	Y
0.455	0.100	1.77
0.500	0.500	2.74
0.500	0.333	2.66
0.500	0.250	2.57
0.600	0.500	5.33
0.600	0.333	4.89

11.6   PAFEC-FE input data

TITLE Fracture Mechanics: Ring Fracture Test
C
C
C   Ring specimen with R_o = 50.0 mm, R_i/R_o = 0.5 and $[a/(R_o-R_i)]$ = 0.5,
C   containing a single crack and subjected to a compressive 2 N load.
C
C
CONTROL
TOLERANCE = 10E–6
CONTROL.END
C
NODES
AXIS = 3

NODE.NUMBER	X	Y
1	0.03750	0.0
2	0.0250	1.0
3	0.050	0.0
4	0.0250	5.0
5	0.03750	4.5
6	0.050	4.5
7	0.0250	9.0
8	0.03750	9.0
9	0.050	9.0
10	0.0250	90.0
11	0.050	90.0
12	0.0250	171.0
13	0.050	171.0
14	0.0250	175.5
15	0.050	175.5
16	0.025	180.0
17	0.050	180.0
21	0.43750	0.50
22	0.081250	0.0
23	0.043750	4.750
24	0.081250	4.50
25	0.043750	9.0
26	0.081250	9.0
31	0.0445312	0.06250
32	0.0804687	0.0
33	0.0445312	0.59375
34	0.03750	0.56250
35	0.0804687	0.56250
36	0.0445312	1.1250
37	0.03750	1.1250
38	0.0804687	1.1250

C

PAFBLOCKS

BLOCK.NUMBER	TYPE	ELEMENT	N1	N2	N3	TOPOLOGY						
1	1	36210	5	3	0	2	1	7	8 0	4	5
2	1	36210	6	3	0	1	3	8	9 0	5	6
3	1	36210	2	1	0	7	9	12	13 0	10	11
4	1	36210	2	3	0	12	13	16	17 0	14	15
5	5	36210	3	3	4	1	21	8	25 0	5	23
6	5	36210	3	3	4	1	22	8	26 0	5	24
7	1	36110	3	3	0	31	1	36	37 0	33	34
8	1	36110	3	3	0	32	1	38	37 0	35	34

C
MESH

REFERENCE	SPACING.LIST				
1	18				
2	0.5	0.5	1	1	1
3	1				
4	1.0	1.5	2.0	2.5	
5	0.5	0.5	1		
6	1	1			

C
REFERENCE.IN.PAFBLOCK
REMOVE = 1

BLOCK.NUMBER	C1	C2	POSITION	NODE
5	1	1	1	1
5	1	1	2	31
5	1	1	3	37
5	1	1	4	36
5	1	1	6	34
5	1	1	7	33
6	1	1	1	1
6	1	1	2	32
6	1	1	3	37
6	1	1	4	38
6	1	1	6	34
6	1	1	7	35
1	3	1	1	21
1	3	1	2	1
1	3	1	3	25
1	3	1	4	8
1	3	1	6	23
1	3	1	7	5
2	1	1	1	1
2	1	1	2	22
2	1	1	3	8
2	1	1	4	26
2	1	1	6	5
2	1	1	7	24

C

```
LOADS
NODE.NUMBER   DIRECTION   VALUE.OF.LOAD
     3            1          −2.0
C
CRACK.TIP
1
C
RESTRAINTS
NODE.NUMBER  PLANE  DIRECTION
     17         4      2
     17         0      1
C
IN.DRAW
TYPE.NUMBER    INFORMATION.NUMBER    SIZE.NUMBER
     2               0                  10
     1               3                  10
END.OF.DATA
```

11.7   Interpretation of PAFEC-FE results

The main goal of a parametric analysis such as the one presented in section 11.5 is to design a ring specimen for which K_I remains constant over a range of non-dimensional crack lengths $[\alpha/(R_o-R_i)]$.

The objective is to specify the details of a specimen that may be used in parametric crack-growth studies, such as experiments utilized in studying the effect of cyclic loading upon crack growth where the influence of changing K_I needs to be excluded.

Examination of the tabulated results presented in section 11.5, reveals that when $R_i/R_o = 0.5$, variation of the non-dimensional crack length from 0.25 to 0.50, appears to have an insignificant effect on the calculated normalized stress intensity factor Y.

However, when R_i/R_o is increased from 0.5 to 0.6, then the non-dimensional crack length $[\alpha/(R_o-R_i)]$ has a prominent influence on the calculated Y. Furthermore, comparison of the Y results for $R_i/R_o = 0.6$ with the corresponding $R_i/R_o = 0.5$ results, shows that as the thickness of the specimen decreases the calculated Y increases. This influence of the ring specimen thickness is further verified by noting that the smallest Y in the tabulated results corresponds to the relatively thicker ring specimen with $R_i/R_o = 0.455$.

Perhaps the most important accomplishment of the parametric analysis presented in section 11.5 is the existence of the excellent agreement with a set of available experimental data. In particular, by making use of the results from tests conducted with ring specimens characterized by $R_i/R_o = 0.5$ and subjected to compressive loading (Jones, 1974), it can be seen from Figure

11.5 that the PAFEC-FE results show a first class correlation with experimental data.

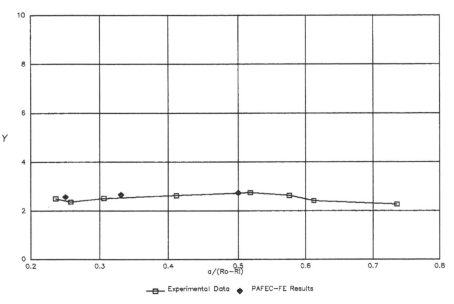

Figure 11.5 Variation of normalized stress intensity factor (Y) with non-dimensional crack length [$a/(R_o - R_i)$] for a ring specimen with $R_i/R_o = 0.5$, containing a radial crack and subjected to compressive diametric loading (comparison of experimental data with PAFEC-FE results).

References

J. Ahmad and N.E. Ashbaugh, Constant K_1 crack-propagation test specimens, *Int. J. Fracture* **19** (1982) 115–129.

H.L. Ewalds and R.J.H. Wanhill, *Fracture Mechanics*, Edward Arnold, London (1985) p. 304.

G.R. Irwin, Analysis of stress and strain near the end of a crack traversing a plate, *J. Appl. Mech.* **24** (3) (1957) 361–364.

A.T. Jones, A radially cracked, cylindrical fracture toughness specimen, *Engineering Fracture Mechanics* **6** (1974) 435–446.

M.F. Kanninen and C.H. Popelar, *Advanced Fracture Mechanics*, Oxford Engineering Science Series, Cambridge University Press, Cambridge (1985).

M.G. Karfakis, K.P. Chong and M.D. Kuruppu, A critical review of fracture toughness testing of rock, in *Proc. 27th US Symp. on Rock Mechanics* ed. H.L. Hartman, SME, Littleton (1986) pp. 3–10.

J.F. Knott, *Fundamentals of Fracture Mechanics*, Butterworths, London (1973).

A. Nadai, *Theory of Flow and Fracture of Solids*, Vol. 1, 2nd edn., McGraw-Hill, New York (1950) p. 572.

J.R. Rice, A path independent integral and the approximate analysis of strain concentration by notches and cracks, *J. Appl. Mech.* **35** (1968) 379–386.

S.P. Shah, ed., *Applications of Fracture Mechanics to Cementitious Composites*, NATO ASI Series E: Applied Sciences Vol. 94, Marinus Nijhoff, Dordrecht (1985).

R.M. Thompson, C.P. Andrasic and A.P. Parkeret, Comparison of two sets of results for radially cracked, point loaded ring specimens, *Engineering Fracture Mechanics* **19** (1984) 383–386.

F.H. Wittman, ed., Fracture mechanics of concrete, in *Developments in Civil Engineering*, Vol. 7, Elsevier Science Publishers, Amsterdam (1983).

Appendices

Appendix A

PAFEC-FE operating notes

To be completed locally:
Computer: Operating system:

Command to run PAFEC-FE: (i)   Interactively on-line
                             (ii) Off line

Command to produce graphics: (i)   On a VDU
                             (ii) On hard copy plotter

Control options which may have local characteristics:

ADD.JC
CLEAR.FILES
CONCATENATE.OUTPUT
FO.SAVE
FULL.OUTPUT
HIGH.SUBS
LIB
MAP
OWN.JC
READ.FROM
SAVE
SAVE.DISPS.TO.<filename>
SAVE.FORTRAN
SAVE.TEMPS
SAVE.TEMPS.TO.<filename>
SCRATCH
SEMI.<filename (s)>
STOP.INCREMENT=<integer>
STORE.INCREMENT=<integer>
TIMER
USE.<filename>

Locally supplied control options:

Index